MEDIA AND COMMUNICATIONS -
TECHNOLOGIES, POLICIES AND CHALLENGES

NOVEL USAGE OF ERBIUM IN OPTICAL COMMUNICATION SYSTEMS: FROM FUNDAMENTALS TO PERFORMANCE CHARACTERISTICS

MEDIA AND COMMUNICATIONS - TECHNOLOGIES, POLICIES AND CHALLENGES

Additional books in this series can be found on Nova's website under the Series tab.

Additional E-books in this series can be found on Nova's website under the E-books tab.

MEDIA AND COMMUNICATIONS -
TECHNOLOGIES, POLICIES AND CHALLENGES

NOVEL USAGE OF ERBIUM IN OPTICAL COMMUNICATION SYSTEMS: FROM FUNDAMENTALS TO PERFORMANCE CHARACTERISTICS

SHANKAR S. PATHMANATHAN
AND
P.K. CHOUDHURY

Nova Science Publishers, Inc.
New York

Copyright © 2010 by Nova Science Publishers, Inc.

All rights reserved. No part of this book may be reproduced, stored in a retrieval system or transmitted in any form or by any means: electronic, electrostatic, magnetic, tape, mechanical photocopying, recording or otherwise without the written permission of the Publisher.

For permission to use material from this book please contact us:
Telephone 631-231-7269; Fax 631-231-8175
Web Site: http://www.novapublishers.com

NOTICE TO THE READER

The Publisher has taken reasonable care in the preparation of this book, but makes no expressed or implied warranty of any kind and assumes no responsibility for any errors or omissions. No liability is assumed for incidental or consequential damages in connection with or arising out of information contained in this book. The Publisher shall not be liable for any special, consequential, or exemplary damages resulting, in whole or in part, from the readers' use of, or reliance upon, this material.

Independent verification should be sought for any data, advice or recommendations contained in this book. In addition, no responsibility is assumed by the publisher for any injury and/or damage to persons or property arising from any methods, products, instructions, ideas or otherwise contained in this publication.

This publication is designed to provide accurate and authoritative information with regard to the subject matter covered herein. It is sold with the clear understanding that the Publisher is not engaged in rendering legal or any other professional services. If legal or any other expert assistance is required, the services of a competent person should be sought. FROM A DECLARATION OF PARTICIPANTS JOINTLY ADOPTED BY A COMMITTEE OF THE AMERICAN BAR ASSOCIATION AND A COMMITTEE OF PUBLISHERS.

LIBRARY OF CONGRESS CATALOGING-IN-PUBLICATION DATA

Pathmanathan, Shankar S.
 Novel usage of erbium in optical communication systems : from fundamentals to performance characteristics / authors, Shankar S. Pathmanathan, P.K. Choudhury.
     p. cm.
 Includes bibliographical references and index.
 ISBN 978-1-61728-955-2 (softcover)
 1. Optical fibers--Materials. 2. Erbium. I. Choudhury, P. K. II. Title.
 TK5103.59.P385 2010
 621.382'75--dc22
                    2010026272

Published by Nova Science Publishers, Inc. ✢ *New York*

Contents

Preface		vii
Chapter 1	Introduction	1
Chapter 2	Numerical Modeling of EDFA Performance Characteristics	9
Chapter 3	Experiments on EDFA under Different Configurations	33
Chapter 4	Summary and Conclusion	49
References		51
Index		55

PREFACE

The twentieth century has witnessed phenomenal growth in silicon-based semiconductor technology. This revolution, however, will be dwarfed by photonics technology in the twenty first century. In this respect, erbium – a rare-earth element on the lanthanide series – will be to photonics technology what silicon is to semiconductor technology. In this article, spotlight shines on the novel usage of erbium in amplifying optical signals – a much desired phenomenon implemented in optical communication systems.

Due to various loss mechanisms, there is a gradual reduction in the power of light as it propagates through a communication channel, and to recover that, amplification of light becomes vital. Further, this remains much important in order for the information carried by the light to be discernible at the receiving end as there exists a minimum threshold power which the light must always possess. Erbium-doped fiber amplifier (EDFA) is an *all-optical* type of device for combating attenuation in optical communication systems wherein amplification performs totally in the optical domain via stimulated emission.

In this article, starting from the fundamental concepts in the context of EDFAs, a thorough description is made incorporating the advancements in the area. In this regard, the basic concepts of modeling the EDFA characteristics remain vital – the technique which involves non-linear rate and propagation equations, the solutions to which are carried out numerically by the use of the forth order Runge-Kutta method. The associated boundary value problem is then solved by using the relaxation method. The results show the important device performance characteristics of gain and noise figure with respect to three main input design parameters – pump power, signal wavelength and input signal power.

The article also involves the experimental investigations of EDFA in the C-band (1530 nm – 1565 nm). In this context, the performance characteristics for various system configurations are illustrated considering different types of pumping schemes viz. co- and counter-propagations of the pump and the signal. The primary performance characteristics of EDFA under different configurations are discussed in terms of gain and noise figure. It has been observed that the most appreciable values of gain are obtained corresponding to the double-pass configuration with a tunable band-pass filter in the optical circuit. However, for a given configuration, pumping schemes are hardly found to play a significant role towards the improvement in gain or reduction in noise figure.

Chapter 1

INTRODUCTION

Light can be utilized to transmit complex information – this gave rise to the fiber-optic communication system, which heralded the arrival of the modern information age – an era of fast and high bandwidth communication. The wonderful property of fiber-optic communication system to guide light is truly marvelous. But, if it is to be realized in practice in a system which spans distances of over thousands of kilometers connecting continents, the shortcoming of signal degradation, owing to its being attenuated by the medium (optical fiber) itself, must be overcome. As such, attenuation remains a serious condition that afflicts light propagating ultra-long distances through a fiber-optic cable. Left unchecked, attenuation results in the gradual loss of light, which leads ultimately to the loss of information. These losses need to be compensated, and is generally accomplished by implementing an amplifier. Erbium-doped fiber amplifier (EDFA), which is an *all-optical* amplifier, is the panacea for this malady called attenuation.

The key element in the EDFA is erbium – a rare-earth element in the lanthanide series. In the past, erbium was a relatively unimportant and unknown element, but the concept has changed dramatically now. In fact, it has been postulated that what silicon is to semiconductor technology, erbium will be to photonics technology. According to Emmanuel Desurvire [1] of Columbia University, the use of erbium doping in small amounts in usual optical fibers "makes it possible to distribute the gain over the fiber itself, thereby minimizing the power excursion of the signal. Such an approach makes possible virtually lossless signal transmission from one fiber network to the next."

In fact, roots of erbium-doped fiber lasers and amplifiers began in 1964 with the first amplification experiment in rare-earth doped fiber lasers. The first laser diode pumped fiber lasers were developed in the year 1974. Therefore, it is only befitting that we take a closer look at the role of erbium in EDFAs. In this article, we first outline the basic mechanisms of loss due to attenuation. Then we describe the operational process of the EDFA, followed by the modeling of this device using an established model by Giles and Desurvire [2] together with the simulation results. In tandem, we also take an experimental approach wherein we present experimental circuits corresponding to various configurations and pumping schemes culminating in a discussion of the experimental results, and their comparison with those through simulations.

1.1. ATTENUATION AND THE ROLE OF EDFA

The capability of fiber-optic communication system to transmit tremendous amount of information over long distance is not without problems. Attenuation, dispersion and nonlinear effects are the three major problems that afflict fiber-optic communication systems [3]. However, in the present article, we emphasize only on the attenuation – the phenomenon of reduction in the optical signal power as it propagates through a fiber-optic cable – given the fact that this is the only issue dealt with in the context of EDFAs.

Light is susceptible to attenuation in a long haul fiber-optic communication system. In order for the information carried by the light to be discernable at the receiving end, there exists a minimum threshold power which the light must always possess. As such, the loss in power due to attenuation needs to be compensated, and this is accomplished by implementing an amplifier, which boosts and maintains power levels above a designated lower limit value. Traditional amplifiers like the much inconvenient electro-optic regenerators, which require high speed electronics dedicated to the specific bit rate, make the overall amplification system rather complex with bulky circuitries involving both the optical as well as the electrical domains. Optical amplifiers are found to be far superior in this regard. Particularly, with the advent of EDFA, which is an all-optical amplifier operating in the C-band, all the thorny difficulties associated with light amplification were easily resolved as these provide several merits over regenerative repeaters as well as other amplification systems. One of the key advantages with EDFAs has been that these do not require high-speed electronics, and the optical gain of the amplifier does not change with bit rate. Such features of EDFAs leave enormous room for expansion into high-speed systems as well as allow the manufacturer to use the same EDFA repeaters for different types of systems with different bit rates.

1.2. OPERATIONAL PRINCIPLE OF EDFA

In order to understand the operational principle of EDFA, we need to review three basic processes – absorption, spontaneous emission and stimulated emission – occurring in the atom. Absorption is a process in which an electron in the lower energy level E_1 absorbs a photon of energy $h\nu$ and makes a transition to a higher energy level E_2. On the other hand, spontaneous emission is a process wherein an electron in a higher energy level E_2 makes a transition to a lower energy level E_1 by emitting a photon of energy $h\nu$. In stimulated emission, an incoming photon of energy $h\nu$ interacts with an electron in a higher energy level E_2 and stimulates it to make a transition to a lower energy level E_1 by emitting a photon of energy $h\nu$. The outcome of the stimulated emission process is two photons – one being the initial incoming photon and the other originating from the emission process when the electron drops down to the lower energy level. The special feature of stimulated emission is that both these photons have identical physical characteristics in respect of frequency, energy, phase, polarization state and the traveling direction. In short, stimulated emission can be considered to be a process that clones photons [4]. Thus, the main processes contributing to the population of higher energy levels are the radiative absorption (decay) and the non-radiative decay; the excitation energy is converted into one or several photons. Given that the stimulated emission replicates photons, the process can be utilized to amplify light. Using a simple model with three energy levels, the process of light amplification (that occurs in an EDFA) can be described in an elementary manner, as depicted schematically in Figure 1 [4].

Firstly, let E_1, E_2 and E_3 represent the energy levels of an erbium ion (Er^{3+}), in the order of increasing energy, and let these energy levels be designated as the ground, metastable and pump states, respectively. Under normal thermal equilibrium conditions, most of the electrons in Er^{3+} ion are in the ground state. In order to excite electrons in the ground state to move up to the pump state, an external energy (in the form of photons of energy $h\nu_{13}$) must be supplied to the system (Figure 1a). In practice, optical pumping is used where an external pump, usually a laser diode, provides this energy. As the pump state has a very short lifetime ($\approx 10^{-9}$ s), the electrons that arrive here, fall down rapidly into the metastable state via non-radiative decay (Figure 1b) – the process in which photons are not emitted. Instead, the excess energy is liberated in the form of quantized lattice vibrations.

Unlike the pump state, the metastable state has a very long lifetime ($\approx 10^{-6}$ s). As a result, electrons falling into the metastable state start to pile up here (Figure 1c). This accumulation of electrons in the metastable state and the almost empty ground state gives rise to a unique condition termed as population inversion – the state in which the number of electrons residing in the metastable level far outnumber the electrons in the ground state (Figure 1c). Having thus created population inversion, any incoming photon of energy $h\nu_{12}$ can interact with the electrons in the metastable state via the stimulated emission process, and trigger the creation of a cascade of stimulated emission photons of energy $h\nu_{12}$ (Figure 1d). In an EDFA, the Er^{3+} ion has a myriad of energy levels, and not all of these energy levels are utilized for stimulated emission. The metastable and the ground states involved in stimulated emission process (in EDFAs) are $^4I_{13/2}$ and $^4I_{15/2}$, respectively.

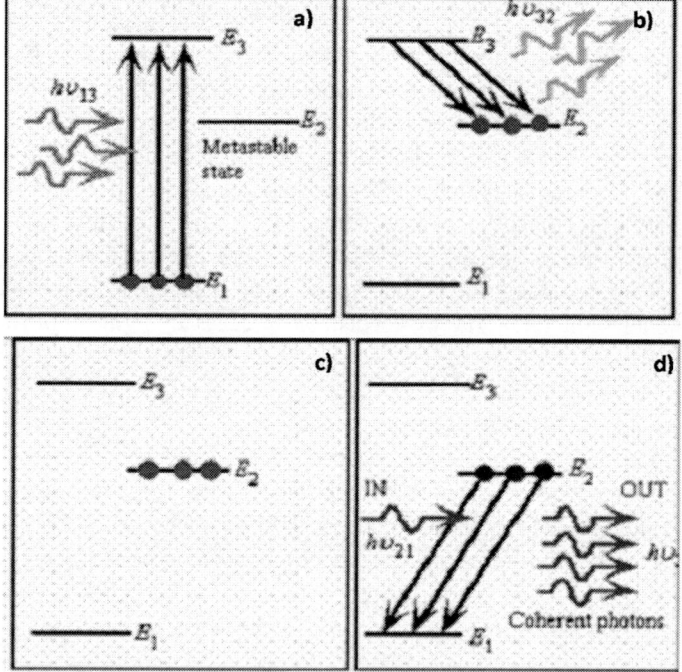

Figure 1. Light amplification process by stimulated emission [4].

1.3. EDFA Development

As stated above, investigations on erbium-doped fiber lasers and amplifiers were started in 1964, followed by the commencement of the first fiber laser in 1974. The EDFA, invented by a team led by David Payne at the University of Southampton, was announced in early 1987. About the same time, another team independently at the AT&T Bell labs led by Emmanuel Desurvire and Randy Giles also achieved the same feat. It was this accomplishment that made possible the global high-capacity optical fiber network, serving as a backbone of the global information superhighway. Progress was rapid; electro-optic regenerators were quickly phased out and replaced by EDFA repeaters in the fourth generation system. Ultimately EDFAs have become indespensible in the 1.55 µm (C-band) region of the electromagnetic spectrum where the fiber loss remains minimum.

A frequently used figure of merit for communication systems is the *bit rate-distance product BL* [5], where B is the bit rate and L is the repeater spacing. It is especially noteworthy that there is a sharp increase, as depicted in Figure 2, in the information-carrying capacity of the system after EDFA repeaters were introduced into the transmission network. A further enhancement in the fourth generation system is the deployment of wavelength division multiplexing (WDM) to increase the bit rate.

The phenomenal growth observed in the telecommunications sector was possible largely due to the intensive research conducted around the globe. The importance of EDFA as the holy grail of optical amplifiers attracted the attention of researchers worldwide. Two parameters – gain (G) and noise figure (NF) – play key roles in the performance of an EDFA, and can be considered as the main figures of merit of the amplifier. Gain determines the power amplification factor of the amplifier, whereas noise figure describes quantitatively the degradation of the signal quality at the output with respect to the signal quality at the input. The higher the gain is, the higher will be the output amplified power. In contrast, a high noise figure indicates a badly degraded amplified signal. The goal, therefore, remains to develop an EDFA with a high gain and low noise figure.

However, there are several issues involved so far as the development of EDFAs is concerned. As stated above, gain and noise figure are the prime factors to determine the EDFA performance. Apart from that, gain saturation also remains one of the important parameters, and all these may be optimized by suitably choosing the amplifier length, the pump absorption band and the pump wavelength within the band, the signal wavelength, the peak Er^{3+} ion concentration profile, and the waveguide characteristics. The decay of pump power along the EDFA length results in a non-uniform inversion of the medium. In fact, the fiber presents lossy characteristic after a certain length because the fiber medium itself absorbs light when it is not inverted. The optimum length of EDFA essentially depends on the pump power since a longer length of inverted medium can be achieved by using a higher pump power. Assuming a fixed signal wavelength, three pump bands near 1480 nm, 980 nm and 810 nm correspond to two- and three-level pumpings. For two-level pumping, the required pump power approaches to infinity when the pump wavelength approaches the signal wavelength, which essentially imposes the need of the pump wavelength to be detuned away from the signal wavelength. However, the absorption coefficient decreases corresponding to shorter wavelength pumps, and this makes higher value of the saturation power, requiring thereby larger pump power for inversion.

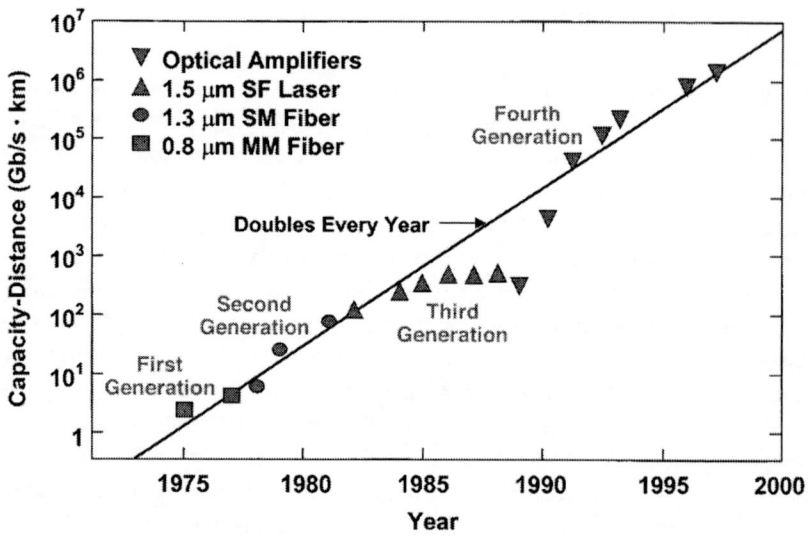

Figure 2. Capacity of lightwave communication systems over several generations [5].

Optimization of waveguide characteristics need numerical aperture, core radius and cutoff wavelength to be suitably considered. Usually diode lasers are used to pump EDFAs, and high gain from a modest amount of pump power essentially requires suitable fiber design and highly efficient pumping conditions. Generally, the Er^{3+} ion concentration profile [6], or the erbium transversal distribution (ETD), is considered to be having a step profile. Nevertheless, the experimentally determined erbium profiles reported by the investigators have relatively smooth wings, and sometime they even have a dip approximately centered at the fiber axis [7–9]. Thus, the ETD essentially has its profound impact on the EDFA performance characteristics in the sense that the same quantity of Er^{3+} ions can yield very different values of gain depending on how they are distributed in the transverse direction. In fact, an adequate choice of ETD profile leads to the EDFA gain optimization together with the EDFA length optimization [10]. As such, aforesaid discussions reveal that the best EDFA performance predicted by design optimization is not just straightforward, and a number of parameters must be taken into consideration during fiber fabrication.

The gain profile of EDFA has been an important aspect, and the gain peak wavelength is determined by estimating the gain per unit length of the fiber [11,12]. In view of this, extensive studies have been performed which comprise of both experimental and theoretical work. Since EDFAs can simultaneously amplify signals over a considerably large bandwidth of about 35 nm, these have become major driving technology for WDM systems in long haul communications [13]. However, EDFAs possess one inherent setback in having non-uniform variation in gain with wavelength yielding thereby a wavelength-dependent amplification characteristic. Nevertheless, techniques for successful gain flattening have been investigated, which include the implementation of long period fiber gratings or Mach-Zehnder filters [14–19], fluoride based EDFAs [20–22] and a suitable co-dopant material [23].

In the context of EDFAs, the research work primarily concentrates on improving the gain and the noise figure by suitably designing optical circuits with various configurations, e.g. single-pass and double-pass configurations as well as incorporating additional components like filters, isolators etc. In this regard, single-pass amplification circuit is the simplest one. Among the other configurations, the techniques implementing double-pass amplification have been reported to be quite efficient to improve the gain and the noise figure characteristics of EDFAs [24–27]. It is to be noted that the EDFA length and the pump power play determining roles to govern the efficiency of long-haul communication systems.

Endeavors to increase the operational bandwidth of the EDFA, which was originally restricted to the C-band, have also borne fruit. There are now EDFAs in the L- and the S-bands too, although these devices are not yet in widespread use. In summary, a few of the notable milestones in the development of EDFA technology are tabulated in Table 1.

The subject matter of the present article concentrates on the device performance characteristics, namely the gain and the noise figure. The focus is made on the performance characteristics of an EDFA of fixed length and specified dopant concentration, which is employed in various system configurations and pumping schemes. Both experimental and computer simulation methods are employed. The efficacy of such an approach lies in that the experimental results generate data that can be implemented to verify through computer simulations. In addition, simulation results can be used as basic building blocks in future constructions of further complex models.

Table 1. Milestones in EDFA research in chronological order

MILESTONES	REFERENCES
First theoretical analysis of EDFA	Urquhart, 1988 [28]
EDFA is demonstrated experimentally	Desurvire *et al.*, 1989 [2]
Mathematical model of EDFA without ASE	Saleh *et al.*, 1990 [29]
Mathematical model of EDFA with ASE	Giles *et al.*, 1991 [30]
EDFA performance improvement using isolator	Yamashita *et al.*, 1992 [31]
EDFA performance improvement using filter	Yu *et al.*, 1993 [32]
An enhanced mathematical model of EDFA	Zech, 1995 [12]
A broad band double pass EDFA	Hwang *et al.*, 2001 [33]
High gain L-band double-pass EDFA	Yi *et al.*, 2004 [34]
EDFA for the S-Band	Thyagarajan *et al.*, 2004 [35]

Chapter 2

NUMERICAL MODELING OF EDFA PERFORMANCE CHARACTERISTICS

Modeling is a purely theoretical approach conceptualized for describing a natural phenomenon in terms of mathematics. It allows one to predict the outcome of a phenomenon in terms of a specified output for a given number of inputs. In a proper perspective, modeling complements experimentation, although experiments form the final basis of verification of an invention. With the advent of powerful computers and robust mathematical software in the late twentieth century, modeling has been catapulted into the forefront of solution techniques, and will continue to play a dominant role in this aspect.

In reality, a wide range of physics is involved in dealing with erbium-doped glass, which makes a comprehensive model difficult. Further, the analysis must evolve with the conditional changes in fiber design, fabrication, and also, the used erbium glass materials. EDFAs based on silica glass were initially treated under the concept of three-level laser system approximations. However, the theory had to be expanded to the case of pump being in the 1480 nm band where the amplifiers actually behave like a quasi-two-level system. Complications arise further with the development of EDFAs based on fluorozirconate glasses as the case involves several excited state levels having long lifetimes, making thereby the two- or three-level models unsuitable to deal with the system.

In a pioneering paper in 1991, Giles and Desurvire [30] published a model for EDFA to determine their performance characteristics. Implementing the concept, in this section, the modeling of EDFA performance characteristics is expounded under the single-pass co-propagating (SP-Co) configuration.

2.1. GILES AND DESURVIRE MODEL

The mathematical model in ref. [30] is taken as the starting point in modeling EDFAs. The model consists of a set of rate equations which describe the temporal population distribution of Er^{3+} ions in the ground level (N_1), metastable level (N_2) and pump level (N_3). In this context, it is worth to make a little description on the model as this is the most fundamental concept widely accepted by the relevant R&D community. In this stream, we now first write the rate equations corresponding to the population distribution in the different energy levels.

The rate equations used in the modeling of EDFAs are as follows [30]:

$$\frac{dN_1}{dt} = -\left[\frac{\sigma_{sa}\Gamma_s}{hv_s A}(P_s + P_a^+ + P_a^-) + \frac{\sigma_{pa}\Gamma_p}{hv_p A}(P_p^+ + P_p^-)\right]N_1 +$$

$$\left[\frac{\sigma_{se}\Gamma_s}{hv_s A}(P_s + P_a^+ + P_a^-) + A_{21} + \frac{\sigma_{pe2}\Gamma_p}{hv_s A}(P_p^+ + P_p^-)\right]N_2$$

$$+ \left[\frac{\sigma_{pe}\Gamma_p}{hv_p A}(P_p^+ + P_p^-)\right]N_3 \qquad (1)$$

$$\frac{dN_2}{dt} = \left[\frac{\sigma_{sa}\Gamma_s}{hv_s A}(P_s + P_a^+ + P_a^-)\right]N_1$$

$$-\left[\frac{\sigma_{se}\Gamma_s}{hv_s A}(P_s + P_a^+ + P_a^-) + A_{21} + \frac{\sigma_{pe2}\Gamma_p}{hv_p A}(P_p^+ + P_p^-)\right]N_2 + A_{32}N_3 \qquad (2)$$

$$N_3 = (N_t - N_1 - N_2) \qquad (3)$$

The model also incorporates a set of propagation equations, as stated below, which describe the spatial evolution of pump, signal, forward amplified spontaneous emission (ASE) and backward ASE powers along the longitudinal length of EDFA.

$$\frac{dP_p^+(z,t)}{dz} = -P_p^+\Gamma_p(\sigma_{pa}N_1 - \sigma_{pe2}N_2 - \sigma_{pe}N_3) - \alpha_p P_p^+ \qquad (4)$$

Table 2. EDFA parameters used in numerical simulations

Parameters	Numerical values
Fiber length, L	14 m
Pump power, P_p	50 mW
Signal power, P_s	0.5 µW = −33 dBm
Total Er^{3+} concentration, N_t	4.12 ×10^{24} ions/m^3
Pump wavelength, λ_p	1480 nm
Signal wavelength, λ_s	1550 nm
Speed of light in vacuum, c	3.0× 0^8 ms^{-1}
Signal frequency, v_s (Hz)	c / λ_s
Pump frequency, v_p (Hz)	c / λ_p
Pump absorption cross-section, σ_{pa}	1.86×10^{-25} m^2
Pump emission cross-section, σ_{pe}	0.42×10^{-25} m^2
Pump emission cross-section (λ_p = 1480 nm), σ_{pe2}	0.42×10^{-25} m^2
Signal emission cross-section, σ_{se}	5.03×10^{-25} m^2
Signal absorption cross-section, σ_{sa}	2.85×10^{-25} m^2
A_{21}	100 s^{-1}
A_{32}	10^{-9} s^{-1}
Signal to core overlap, Γ_s	0.4
Pump to core overlap, Γ_p	0.4
Planck's constant, h	6.626×10^{-34} Js
Speed of light in vacuum, c	3.0×10^8 ms^{-1}
Background loss at signal wavelength, α_s	0 m^{-1}
Background loss at pump wavelength, α_p	0 m^{-1}
Area, A	12.6×10^{-12} m^2

$$\frac{dP_p^-(z,t)}{dz} = P_p^- \Gamma_p \left(\sigma_{pa} N_1 - \sigma_{pe2} N_2 - \sigma_{pe} N_3 \right) + \alpha_p P_p^- \tag{5}$$

$$\frac{dP_s(z,t)}{dz} = P_s \Gamma_s \left(\sigma_{se} N_2 - \sigma_{sa} N_2 \right) - \alpha_s P_s \tag{6}$$

$$\frac{dP_a^+(z,t)}{dz} = P_a^+ \Gamma_s \left(\sigma_{se} N_2 - \sigma_{sa} N_1 \right) + 2\sigma_{se} N_2 \Gamma_s h v_s \Delta v - \alpha_s P_a^+ \tag{7}$$

$$\frac{dP_a^-(z,t)}{dz} = -P_a^-\Gamma_s(\sigma_{se}N_2 - \sigma_{sa}N_1) - 2\sigma_{se}N_2\Gamma_s h\nu_s \Delta\nu + \alpha_s P_a^- \qquad (8)$$

In these equations, P_s represents signal power whereas P_p^+ and P_a^+, respectively, are the pump and the forward ASE powers co-propagating with the signal. P_p^- and P_a^- represent, respectively, the pump and the backward ASE powers counter-propagating with respect to the signal. The rest of the symbols are defined in Table 2.

2.2. MODELING OF EDFA WITHOUT ASE

In modeling EDFA performance characteristics, the ideal case of co-propagating pump and signal powers is considered in an EDFA, in the absence of ASE. The utility of this simplest case allows one to model, investigate and interpret all the key characteristics of an EDFA in a much easier and efficient manner. For this purpose, eqs. (1), (2), (5) and (8) are used, whereby the parameters P_p^-, P_a^+ and P_a^- are set to zero. The resulting equations are then manipulated and arranged into the following form:

$$\frac{dN_1}{dt} = -A_0 N_1 + B_0 N_2 + C_0 \qquad (9)$$

$$\frac{dN_2}{dt} = D_0 N_1 - E_0 N_2 + F_0 \qquad (10)$$

where the coefficients A_0, B_0, C_0, D_0, E_0 and F_0 are defined as follows:

$$A_0 = \frac{\sigma_{sa}\Gamma_s P_s}{h\nu_s A} + \frac{\sigma_{pa}\Gamma_p P_p^+}{h\nu_p A} + \frac{\sigma_{pe}\Gamma_p P_p^+}{h\nu_p A} \qquad (11)$$

$$B_0 = \frac{\sigma_{se}\Gamma_s P_s}{h\nu_s A} + A_{21} + \frac{\sigma_{pe2}\Gamma_p P_p^+}{h\nu_s A} - \frac{\sigma_{pe}\Gamma_p P_p^+}{h\nu_p A} \qquad (12)$$

$$C_0 = \frac{\sigma_{pe}\Gamma_p P_p^+ N_t}{h v_p A} \qquad (13)$$

$$D_0 = \frac{\sigma_{sa}\Gamma_s P_s}{h v_s A} - A_{32} \qquad (14)$$

$$E_0 = \frac{\sigma_{se}\Gamma_s P_s}{h v_s A} + A_{21} + \frac{\sigma_{pe2}\Gamma_p P_p^+}{h v_p A} + A_{32} \qquad (15)$$

$$F_0 = A_{32} N_t \qquad (16)$$

When the steady state (dynamic equilibrium) is achieved, the population in each of the energy levels is fixed and remains constant, and hence, can be computed. This means, in steady state, the time derivatives of N_1 and N_2 are zero, i.e.

$$\frac{dN_1}{dt} = \frac{dN_2}{dt} = 0 \qquad (17)$$

This indicates that the left hand side of eqs. (9) and (10) are set to zero. The ensuing simultaneous equations can then be solved. Thus, N_1 and N_2 can be expressed in terms of the coefficients A_0, B_0, C_0, D_0, E_0 and F_0 as follows:

$$N_1 = \frac{C_0 E_0 + B_0 F_0}{A_0 E_0 - B_0 D_0} \qquad (18)$$

$$N_2 = \frac{C_0 D_0 + A_0 F_0}{A_0 E_0 - B_0 D_0} \qquad (19)$$

Next, eqs. (18) and (19) are substituted into the propagation eqs. (4) and (6) – a set of coupled differential equations that cannot be solved analytically. Thus, there is a need to resort to a suitable numerical technique. The standard Runge-Kutta fourth order (RK4) method [36] remains the most suitable numerical technique to be utilized in this case because of its simplicity, good accuracy, ease in programming and good stability. The application software used in carrying out the simulation is MATLAB; Gilat [37] gives an excellent introduction to this software.

It is to be emphasized here that the numerical modeling of EDFA must be based on a three-level system, and should not be reduced to a two-level one, although the population in the pump level (N_3) is very much smaller (negligible) compared to the population in the ground level (N_1) and metastable level (N_2). The reason is – in steady state, no population inversion can be achieved in a true two-level system (for an optical gain), no matter how the system is pumped [38].

The modeling of EDFA in the absence of ASE is an initial-value problem (illustrated schematically in Figure 3) where all the conditions are specified at a single point (in this case at $z = 0$), and the numerical solution is easily achieved by applying the RK4 method. Firstly, we consider a simple case of an ideal EDFA with no background loss in the absence of ASE with suitable physical parameters. The gain (G) and the noise figure (NF) are given by Zervas [39] as follows:

$$G = 10\log_{10}\left(\frac{P_{out}}{P_{in}}\right) \qquad (20)$$

$$NF(dB) = 10\log_{10}(NF) \quad \text{with} \quad NF = \frac{1}{G}\left[\frac{P_{ASE(+)}}{h\nu\Delta\nu} + 1\right] \qquad (21)$$

Here $P_{ASE(+)}$ is the forward traveling ASE for both orthogonal polarizations at the output of the amplifier, G is the net gain measured in decibels, h is Planck's constant, ν is the signal frequency and NF (dB) is the noise figure in decibels.

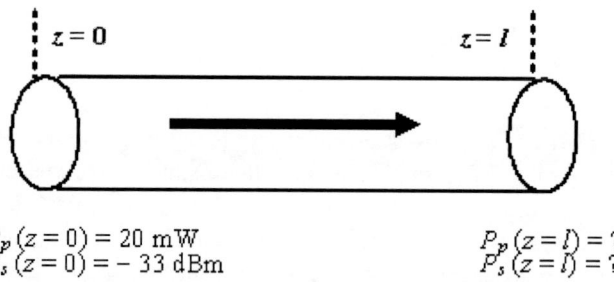

Figure 3. Modeling of EDFA without ASE (an initial-value problem).

2.2.1. Results for EDFA without ASE

Corresponding to a 60 m length of fiber, the variation in gain (dB) along the longitudinal length of EDFA is shown in Figure 4. We observe that, as the longitudinal length (of the fiber) increases, the gain shows a sharp increase till a maximum point is reached, after which the gain begins to drop gradually. This maximum gain corresponds to the optimum fiber length – the length in which there occurs a net growth (increase) in the number of signal photons. Beyond the optimum fiber length, this net growth in gain ceases; instead, a net decay (decrease) of signal photons takes place which causes the gain to drop. Thus, the EDF beyond the optimum fiber length acts effectively as a strong attenuating medium (due to Rayleigh scattering by Er^{3+} ions in silica matrix), causing thereby to reduce the gain. The optimum fiber length in our simulation is observed to be about 39.2 m.

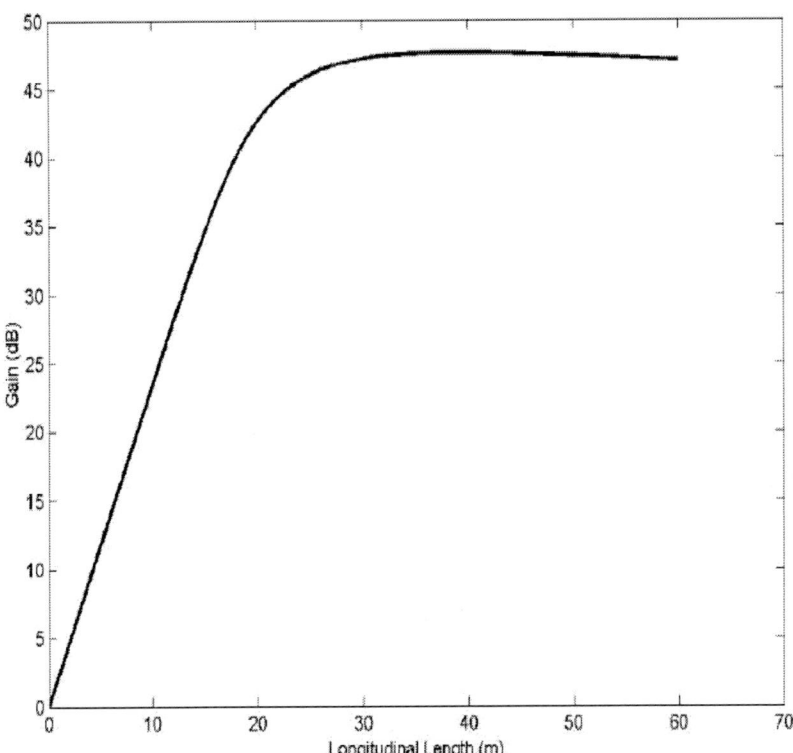

Figure 4. Plot of gain (dB) vs. longitudinal length of EDFA.

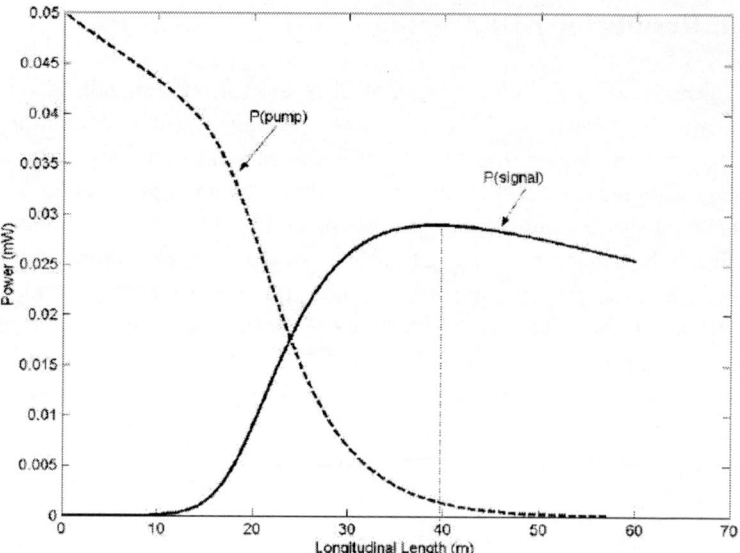

Figure 5. Evolution of pump and signal powers with the longitudinal length of EDFA.

The evolution of pump and signal powers along the longitudinal length of EDFA is shown in Figure 5. We observe that the pump power decreases with the increase in fiber length because Er^{3+} ions in the ground level absorb the pump energy to get excited to the pump level before decaying (non-radiatively) quickly to the metastable level. The pump energy is thereby converted to form and maintain the population inversion. In the presence of population inversion, an incoming signal photon can produce a cascade of stimulated emission photons as it passes through this population inverted region. Thus, a multiplication in the number of signal photons is effectively achieved, causing thereby an increase in the signal power. From the propagation equation

$$\frac{dP_s}{dz} = P_s \Gamma_s (\sigma_{se} N_2 - \sigma_{sa} N_1) - \alpha_s P_s ,$$

assuming $\alpha_s = 0$ (as per our simulation), P_s can grow if and only if

$$(\sigma_{se} N_2 - \sigma_{sa} N_1) > 0 \qquad \text{or} \qquad N_2 > \left(\frac{\sigma_{sa}}{\sigma_{se}}\right) N_1 ,$$

which (in our simulation) is $N_2 > 0.57 N_1$. A plot of $\sigma_{sa}N_1$ and $\sigma_{se}N_2$ along the longitudinal length of EDFA is stated in Figure 6. It is observed that $\sigma_{se}N_2 > \sigma_{sa}N_1$, from the zero length of fiber till its optimum length of 39.2 m.

A plot of N_1 and N_2 along the longitudinal length of the fiber is illustrated in Figure 7. It is noticed that the length, at which the phenomenon of population inversion ($N_2 > N_1$) exists, is from zero till a fiber length of approximately 22.3 m, which is shorter than the optimum fiber length of 39.2 m. The region AB in Figure 7 is to be put in careful attention owing to the fact that, although no population inversion ($N_2 < N_1$) exists in this region, there is yet a net growth or increase in the number of signal photons. This is due to the criteria for signal growth still being fulfilled, here namely $(\sigma_{se}N_2 - \sigma_{sa}N_1) > 0$, because the signal emission cross-section (σ_{se}) is larger than the signal absorption cross-section (σ_{sa}).

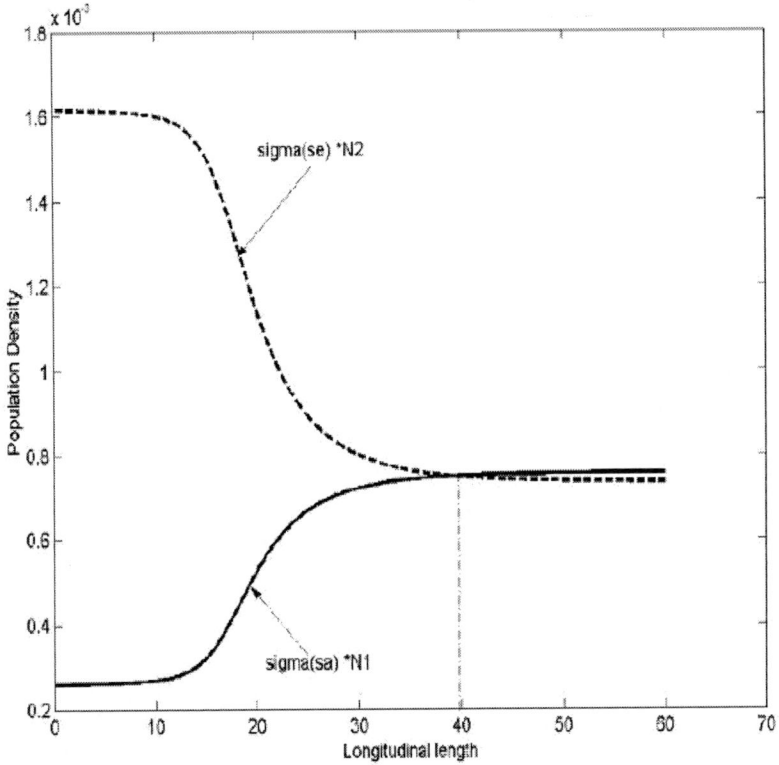

Figure 6. Evolution of population levels with the longitudinal length of EDFA.

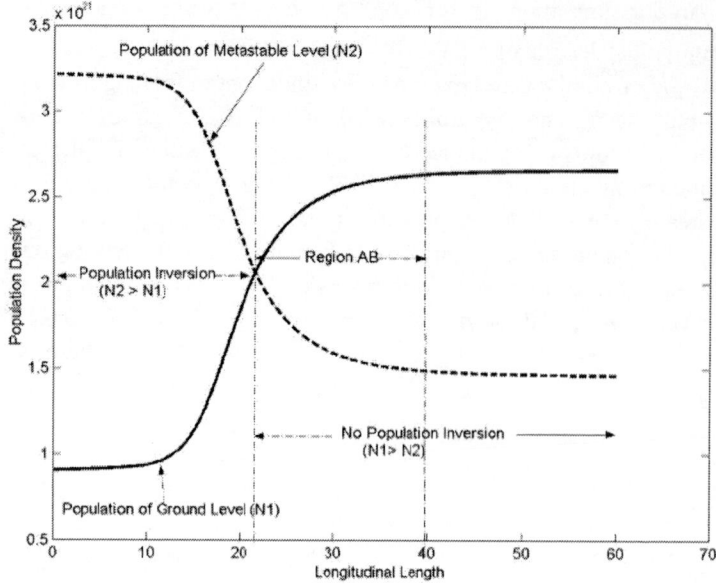

Figure 7. Population density of metastable and ground levels across the longitudinal length of EDFA.

2.3. MODELING OF EDFA WITH ASE

Till now we discussed the EDFA modeling under the ideal situation without considering the effect of ASE. However, the phenomenon of ASE always has its dominant role in signal propagation through EDFAs. As such, taking into account the occurrence of ASE will effectively provide a more realistic approach – a boundary-value problem (Figure 8), whereby all the conditions are not specified at a single point in the domain, but instead, are stated at two or more different points across the domain. In general, the quest for a solution to a boundary-value problem is more intricate and challenging. The solution to a boundary-value problem can be divided into three categories – having no solution, having a single and unique solution or having infinite number of solutions. Normally, the nature of its solution is unknown beforehand.

In the case of EDFA with ASE, the fact that such a device has been realized practically in the laboratory suggests strongly a single and unique solution to this boundary-value problem. Therefore, we can proceed to find the solution by using the standard RK4 method to carry out the forward and happens due to the fact that a complete population inversion is achieved in the fiber, and the population of Er^{3+} ions in the metastable level is now limited by the total Er^{3+} ion concentration in the fiber. On the other hand, the noise figure decreases monotonically as the pump power increases (Figure 12). A high amount of pump power enables a much more complete population inversion to be created across the entire fiber length, reducing thereby the spontaneous emission factor (n_{sp}) which contributes to the reduction in the noise figure.

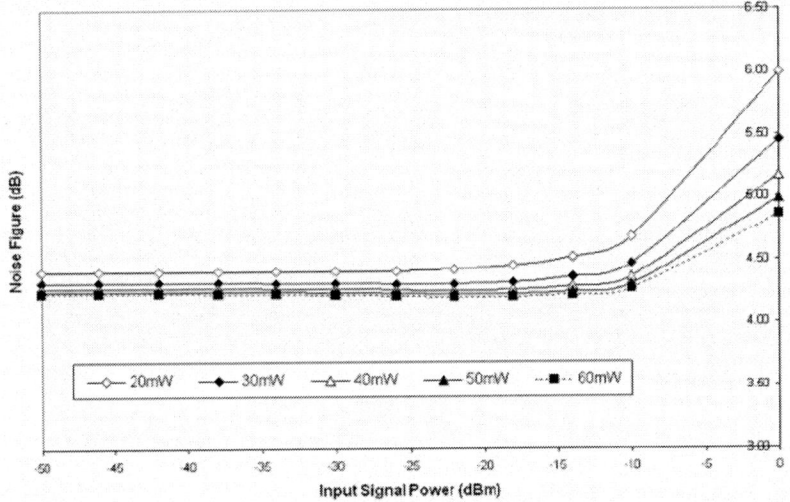

Figure 14. Noise figure vs. input signal power for SP-Co configuration for various pump powers.

The simulation results corresponding to the gain and the noise figure, as a function of input signal power, for various pump powers are depicted in Figures 13 and 14, respectively. For a small input signal power, which is normally in the range of −50 dBm to −30 dBm (designated as the small signal range), we notice that the gain remains fairly constant for a fixed pump power and fiber length. As the input signal power increases beyond −30 dBm, the gain gradually begins to drop linearly. This region of large signal power is designated as the gain saturation region. Here, enormous input signal photons deplete the number of Er^{3+} ions in the metastable level that form population inversion. As a result, the population inversion is reduced, and the gain is curtailed by the limited number of Er^{3+} ions in the metastable level forming population inversion. Likewise, the noise figure shows similar characteristics (Figure 14) of remaining constant in the small signal region for a fixed pump power and EDFA length. However, beyond an input signal power of −10 dBm, the noise figure shoots up. Noting that the noise figure is inversely proportional to the gain, and the gain is decreasing in this region, this behavior of the noise figure is very much anticipated.

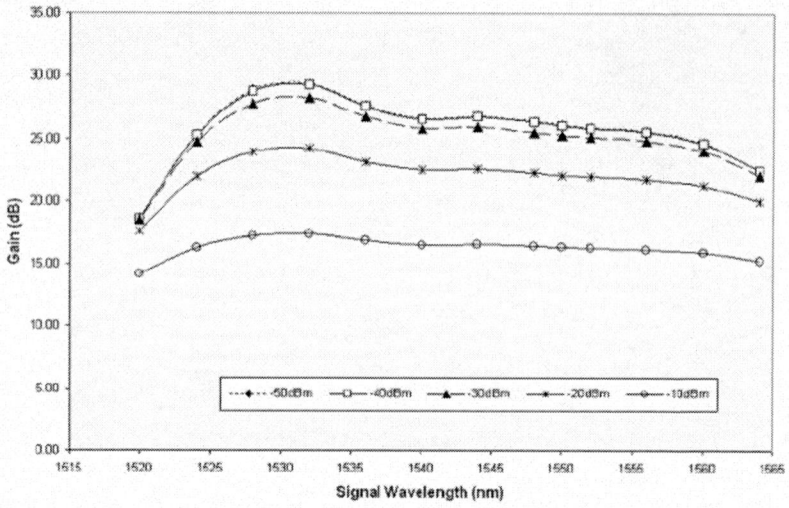

Figure 15. Gain vs. signal wavelength for various input signal powers in SP-Co configuration at a fixed pump power of 20 mW.

Till now we discussed about the variation of EDFA gain and noise figure corresponding to various values of pump power. These features are also observed for their variations against the input signal power. However, besides the above characteristics, the gain spectrum too remains another important feature of the EDFA. In general, the dependence of gain on signal wavelength is termed as the gain spectrum. In our simulation, we limit our signal wavelength to the C-band (i.e. 1520 nm − 1564 nm) only. The gain spectrum and the corresponding variation of noise figure with respect to signal wavelength for different values of input signal power in a SP-Co configuration are illustrated in Figures 15 and 16, respectively. In our simulation process, we keep the pump power constant at 20 mW. We observe that the characteristic curves of gain spectrum (Figure 15) illustrate a strong dependence of gain on signal wavelength. When a small input signal power (−50 dBm to −30 dBm) is used, the dependence is prominent, but almost vanishes in the presence of a large input signal (−10 dBm).

The intrinsic features of gain spectrum are strongly influenced by the atomic properties of the host material into which Er^{3+} ions are doped, namely the absorption cross-section (σ_{sa}) and the emission cross-section (σ_{se}), both of which are wavelength-dependent parameters. The cross-section of a process is a quantitative measure of the probability that the particular process will occur. A larger value of cross-section translates into a bigger likelihood that the process will take place. The absorption and the emission spectra for Er^{3+} ion (Figure 10) indicate that, at shorter wavelengths, the absorption cross-section is larger than the emission cross-section. This makes the absorption as the dominant process, reducing thereby the gain. On the other hand, at larger wavelengths, the emission cross-section is larger than the absorption cross-section, resulting thereby the emission process to be dominant, and simultaneously facilitating the growth of signal gain.

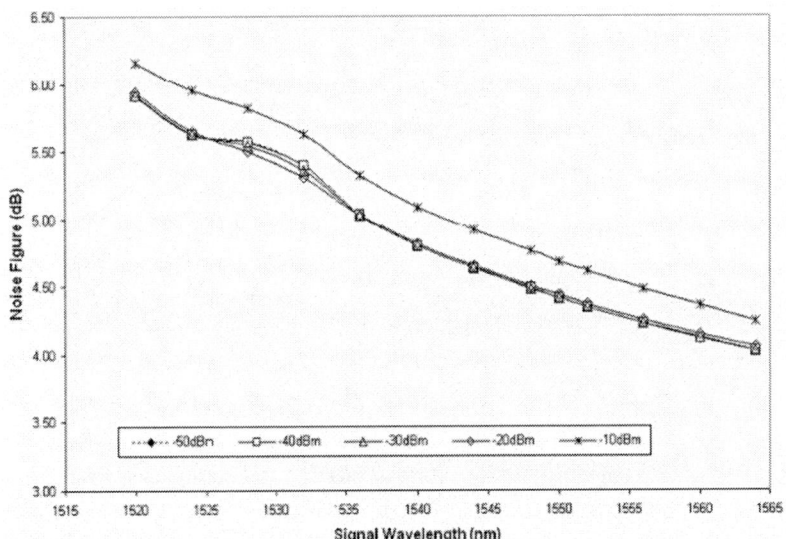

Figure 16. Noise figure vs. signal wavelength for various input signal powers in SP-Co configuration at a fixed pump power of 20 mW.

For a small input signal, which constitutes a signal-photon-limited system, the traits observed in the gain spectrum will be heavily reliant on the probabilistic nature of the emission cross-section. However, in the case of a large input signal, where the number of signal photons injected into the fiber is large, considerable depletion of Er^{3+} ions in the metastable state occurs. This phenomenon reduces the population inversion significantly. In such a system, where the number of signal photons is huge and the population inversion is limited, the influence of small emission cross-section on the gain is negligible. This is owing to the fact that the large flux density of signal photons ensures that one of the many signal photons will ultimately interact with the excited Er^{3+} ion, thus ensuring that the signal amplification occurs.

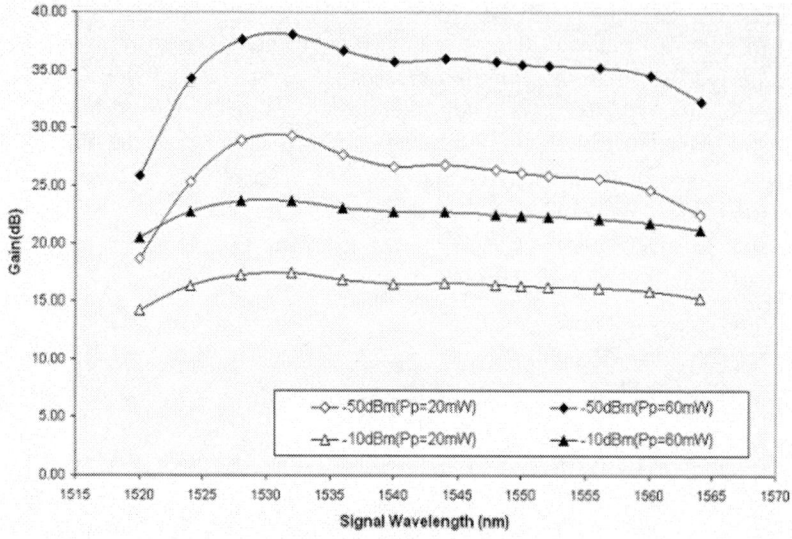

Figure 17. Gain vs. signal wavelength in SP-Co configuration for various values of small and large input signal powers and pump powers.

In simpler terms, a process 'A' with a high probability (analogous to a large emission cross-section) will need only a few trials (analogous to a small number of signal photons) for the process to occur, whereas a process 'B' with a small probability (analogous to a small emission cross-section) will require many more trials (analogous to a large number of signal photons) before it occurs. In this sense, the high input signal flux density possesses the property to offset the low emission cross-section, and therefore, leads to the gain-flattening feature. This is as observed across the C-band in the gain spectrum for high input signal power. In Figure 16, we notice that the noise figure decreases as the signal wavelength increases. The noise figure for a small input signal is smaller than that for a large input signal.

Figure 17 illustrates the gain spectrum for a combination of extremes of large and small input signal power and pump power in a SP-Co configuration. We observe that the effect of large and small input signal powers on the gain spectrum is to alter the characteristic profile of the curve, and has been discussed above. On the other hand, by varying the pump power, the characteristic profile of the gain spectrum does not show much change, but it serves to only translate the entire curve up to a higher gain or down to a lower gain, depending on whether a high pump power (60 mW) or low pump power (20 mW) is utilized for a fixed value of input signal power.

The noise figure characteristic (Figure 18) shows that, for a high pump power, the corresponding noise figure is lower than that observed for a low pump power and a fixed input signal power. A high pump power creates a much more complete population inversion in the fiber, allowing thereby a higher number of fractional stimulated emission process to occur. This reduces the number of fractional spontaneous emission process, and produces a lower amplified spontaneous emission (ASE), reducing thereby the noise figure. We observe from Figure 18 that the noise figure decreases monotonically as the wavelength increases.

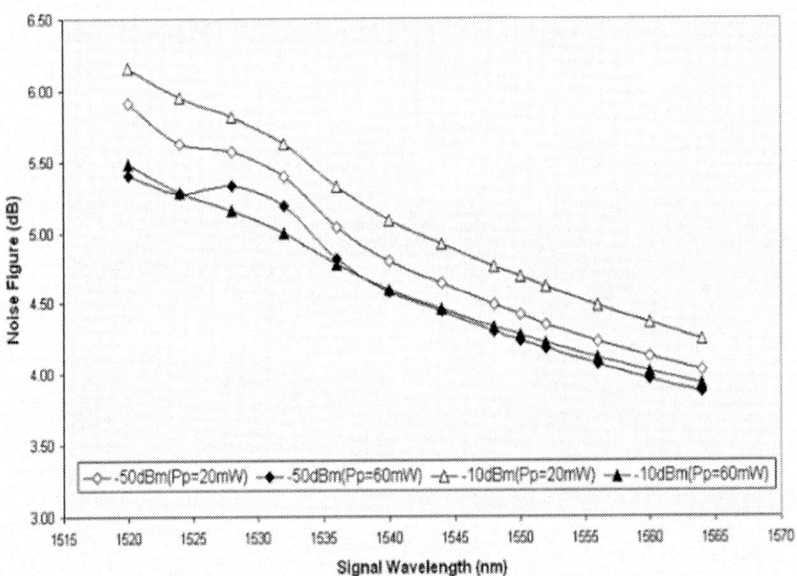

Figure 18. Noise figure vs. signal wavelength in SP-Co configuration for various values of small and large input signal powers and pump powers.

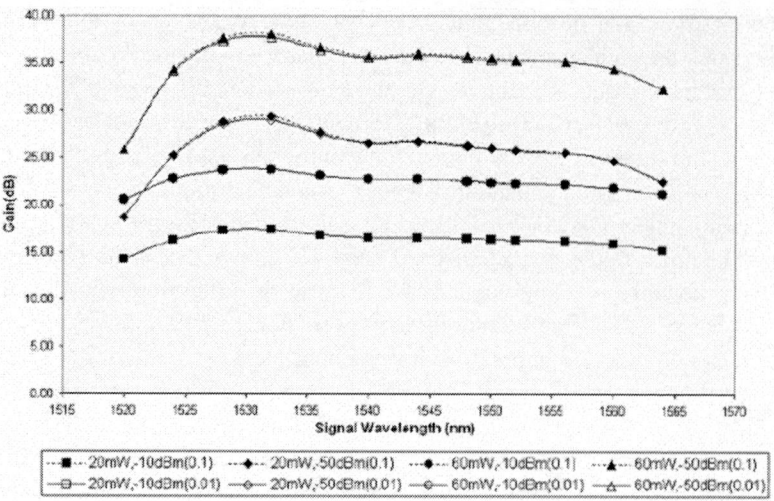

Figure 19. Gain spectra in SP-Co configuration for various values of small and large pump powers and input signal powers at different resolutions (step sizes of 0.1 and 0.01).

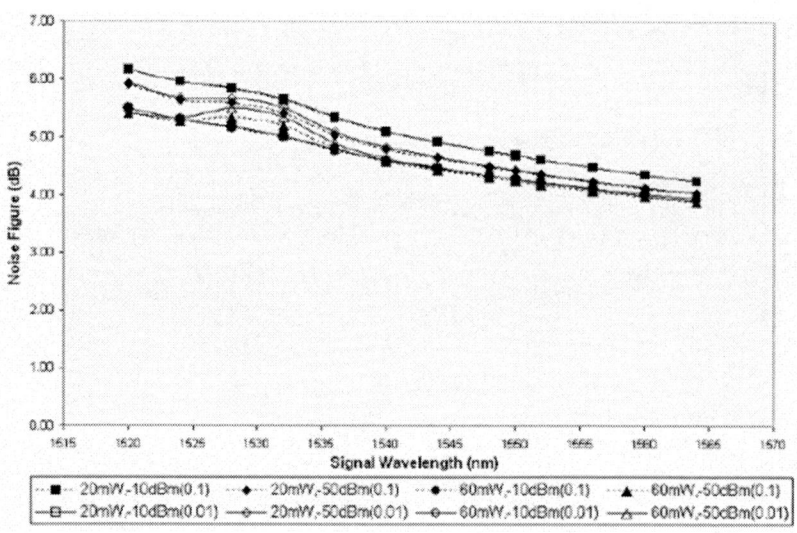

Figure 20. Noise figure vs. signal wavelength in SP-Co configuration for various values of small and large pump powers and input signal powers at different resolutions (step sizes of 0.1 and 0.01).

Figures 19 and 20, respectively, present the gain spectrum and the corresponding wavelength-dependent noise figure characteristics for various combinations of large and small input signal powers and pump powers in the case of a SP-Co configuration for different numerical resolutions (defined as the partitioning of the domain of integration into a number of small, equally-sized segments). We observe that, in comparison to a step size of 0.1, use of a higher numerical resolution (step size 0.01) produces no distinct improvement over the former. Instead, a similar result is obtained, and this justifies the choice of using a lower step size of 0.1 to reduce the computational time without sacrificing the accuracy of results.

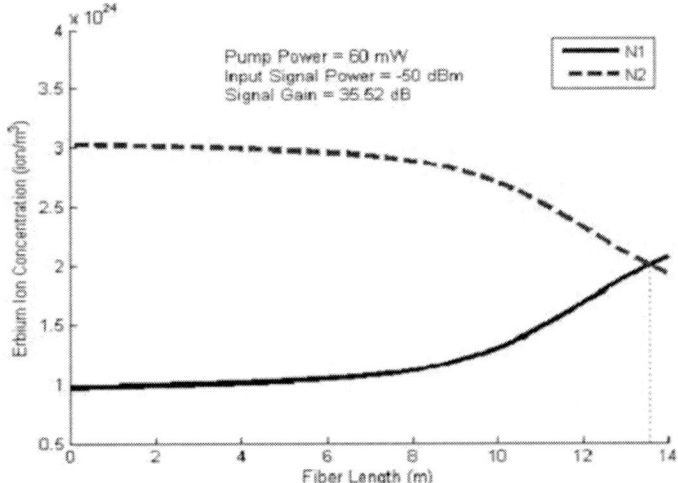

Figure 21. Population distribution of ground and metastable states for a pump power of 60 mW and input signal power of −50 dBm.

As the population inversion remains a vital phenomenon in the operation of EDFAs, a simulation of the population distribution of Er^{3+} ions in the different energy levels, namely the ground state (N_1) and the metastable state (N_2), is essentially an important factor in determining the gain. The greater the population of Er^{3+} ions in the metastable state across the longitudinal fiber length, the larger the gain is. In Figures 21 and 22, the input signal power is small (−50 dBm), and therefore, the gain is constrained by the pump power. The higher the pump power is, the larger the gain becomes until it reaches a particular pump power (60 mW) when gain saturation occurs − the situation when nearly the entire longitudinal length of the fiber is completely inverted. In Figures 23 and 24, the input signal power is large (−10 dBm), and therefore, the gain is constrained by the signal power. The high input signal power reduces greatly the population inversion across the longitudinal length of the fiber, resulting thereby in a lower gain. Finally, Figures 21, 22, 23 and 24 demonstrate that, for a given fiber of fixed length and Er^{3+} ion concentration, the greater the longitudinal length of EDFA (in which a population inversion is set up), the higher the gain will be for that particular EDFA.

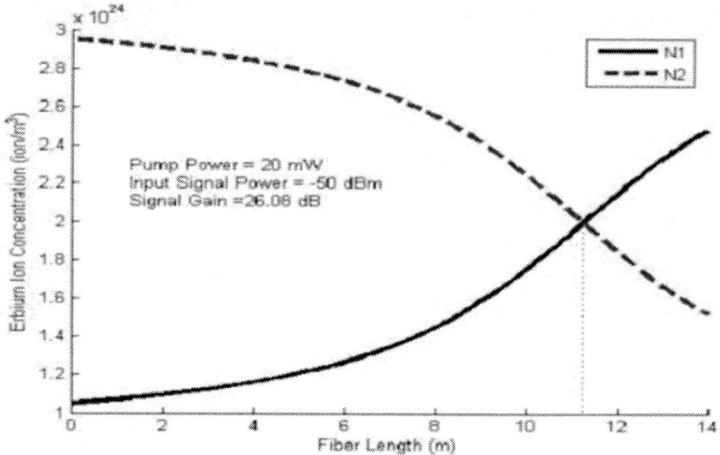

Figure 22. Population distribution of ground and metastable states for a pump power of 20 mW and input signal power of −50 dBm.

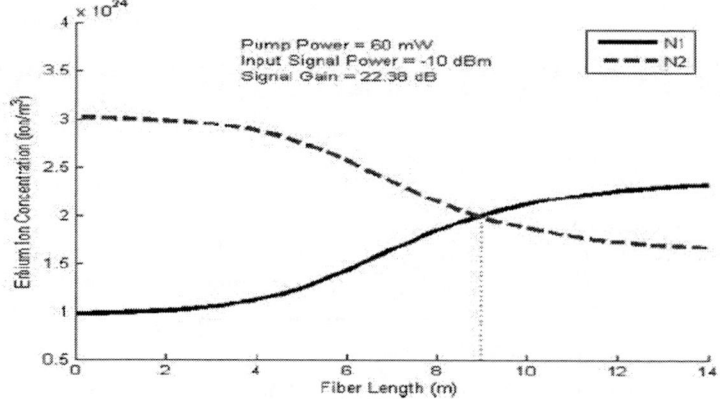

Figure 23. Population distribution of ground and metastabe states for a pump power of 60 mW and input signal power of −10 dBm.

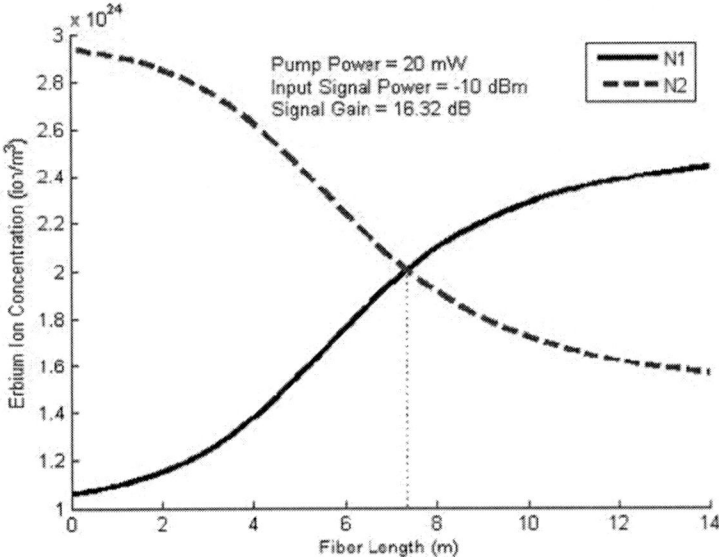

Figure 24. Population distribution of ground and metastable states for a pump power of 20 mW and input signal power of −10 dBm

Chapter 3

EXPERIMENTS ON EDFA UNDER DIFFERENT CONFIGURATIONS

In this chapter, we give an account of the experimental investigations in respect of EDFA characteristics under single-pass and double-pass configurations. The most fundamental optical circuit for EDFA remains as the single-pass configuration. After discussions on the results corresponding to this basic configuration, we touch upon the cases of double-pass EDFA with and without a tunable band-pass filter (TBF) implemented in the circuit. The use of TBF is essentially for the elimination of the unwanted ASE whereby the noise is introduced. In our discussions, we primarily consider two cases – (i) EDFA utilized in a standard double-pass configuration without a filter in which the input signal traverses the gain medium twice and (ii) EDFA employed in an optical circuit which incorporates an optical filter, specifically a TBF, in a standard double-pass configuration. The experimental setups are presented whereby the optical circuit components are identified and their respective functions are elucidated. The working essentials of all the circuits are also described. Following this, the results obtained experimentally are presented, and discussions are made in respect of the comparative characteristics of these optical circuits under various configurations.

3.1. EXPERIMENTAL SETUP

First of all, we study the EDFA characteristics in its most fundamental form – the single-pass configuration. As the name suggests, in a single-pass configuration, the signal traverses the amplifying medium (i.e. the EDFA) only once. The EDFA in a double-pass configuration is an extension of the case corresponding to the single-pass configuration where the signal traverses the amplifying medium twice. Besides that, there are two pumping schemes that we consider in our experiment – co-propagating (or co-pumping) and counter-propagating (or counter-pumping) schemes. In the former case, the pump light is injected in the same direction of travel as that of the signal light, whereas the latter scheme considers the pump and the signal lights traveling in the opposing directions. Optical circuits for SP-Co and single-pass counter-propagating (SP-Counter) configurations are illustrated in Figures 25 and 26, respectively. Also, the optical circuits for double-pass co-propagating (DP-Co) and double-pass counter-propagating (DP-Counter) configurations are shown in Figures 27 and 28, respectively.

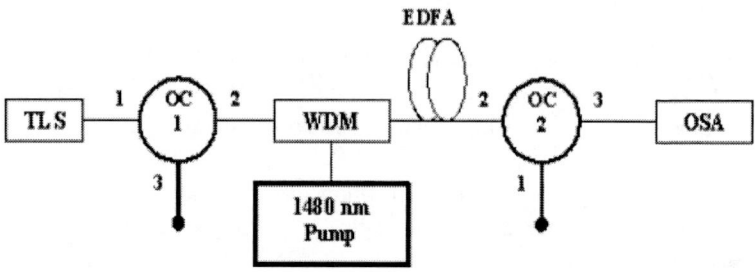

Figure 25. SP-Co system configuration.

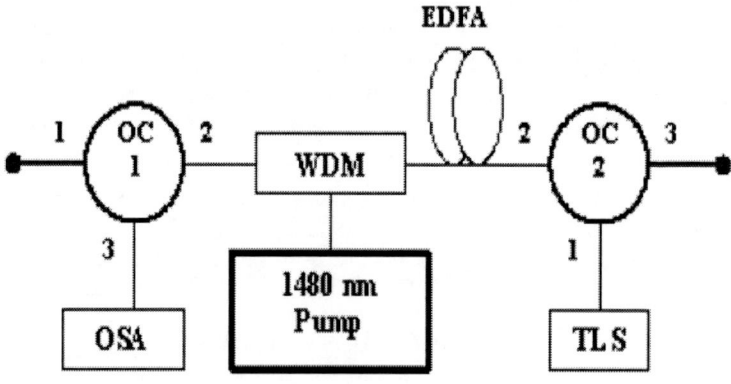

Figure 26. SP-Counter system configuration.

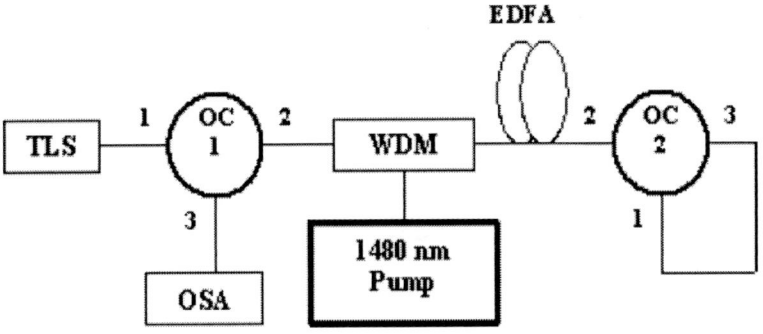

Figure 27. DP-Co system configuration.

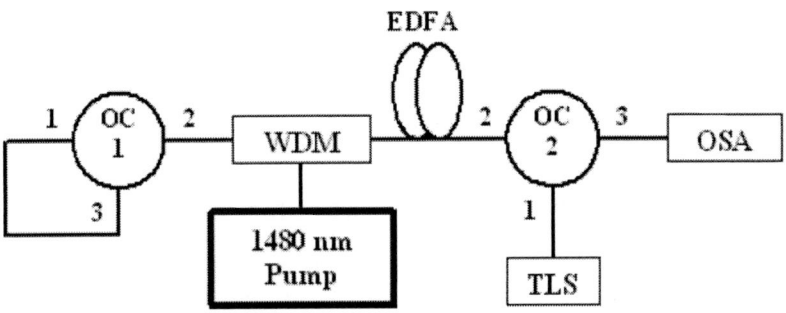

Figure 28. DP-Counter system configuration.

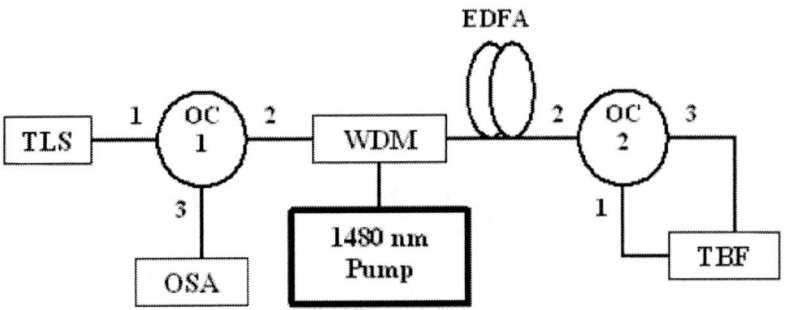

Figure 29. DP-Co-TBF system configuration.

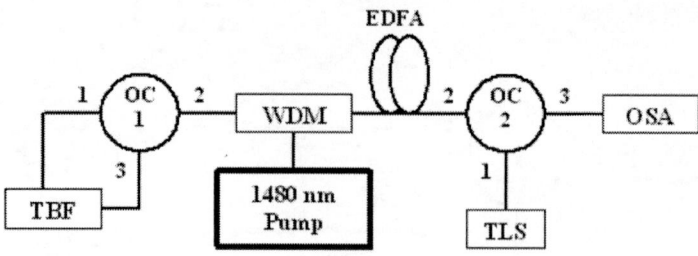

Figure 30. DP-Counter-TBF system configuration.

The EDFA in a double-pass configuration is purposely introduced with a TBF too [26] in the optical circuit to filter out the optical noise originating from the broadband ASE in the input signal power produced during the first pass before it re-enters the second pass for the consecutive amplification process. Optical circuits for double-pass co-propagating with implemented TBF (DP-Co-TBF) and double-pass counter-propagating with TBF (DP-Counter-TBF) configurations are shown in Figures 29 and 30, respectively.

As to the specifications of the EDFA used in our experiment, it is having a length of 14 meter with a concentration of 440 ppm. The EDFA is pumped by a 1480 nm laser diode. A tunable laser source (TLS) is used as the signal source, and the signal wavelength is varied across the C-Band (1528 nm – 1564 nm). In order to regulate the input signal power (measured in dBm), the TLS is connected to a variable-optical attenuator (not shown in the illustrated diagrams). A 1480/1550 nm wavelength division multiplexer (WDM) combines the pump power and the input signal power. C-band three-port optical circulators OC1 and OC2 function as both optical isolators that allow only unidirectional flow of light power as well as routers to guide light into the required fiber. A C-band TBF (with a 3-dB bandwidth of 0.5 nm) is used to filter out the optical noise. The output signal is measured by using an optical spectrum analyzer (OSA). The removal of optical circulators OC1 and OC2 for single-pass configuration is possible provided they are replaced by optical isolators.

In this experimental work [26,27], we investigate the characteristics of the device performance parameters (gain and noise figure) with respect to three different input parameters – the pump power, the signal wavelength and the input signal power. Firstly, for an EDFA of a specified length and Er^{3+} ion concentration, it is important to know the threshold pump power required to saturate the amplifier gain. This is so because pumping beyond the threshold pump power does not elevate the gain significantly, and is merely a waste of pump energy. Next, amplification of the input signal by EDFA is wavelength-dependent, and therefore, the degree of amplification for different wavelengths in the C-Band varies. Thus, the gain changes substantially across the C-band. Knowledge of this behavior is important, especially in systems which amplify multiple wavelength signals using a single optical amplifier. Lastly, the input signal power also influences the performance parameters and the efficiency of the EDFA. In our study, the domain of the input signal power spans from −50 dBm (small signal) to −10 dBm (large signal), and we restrict our wavelength domain to the C-band, which is the operational bandwidth of the EDFA.

3.2. EXPERIMENTAL RESULTS

The gain-pump power curves for a fixed signal wavelength, $\lambda_s = 1550$ nm and a small input signal power ($P_s = -50$ dBm) for SP-Co, SP-Counter, DP-Co and DP-Counter configurations are depicted in Figure 31. In the small signal regime, the gain increases tremendously for a small increase in pump power. As the pump power increases, so does the gain, until it begins to level off. This plateau region indicates that the gain saturation has occurred and the EDFA is (almost) completely inverted. The value of optimum pump power to achieve gain saturation for all configurations is approximately 60 mW.

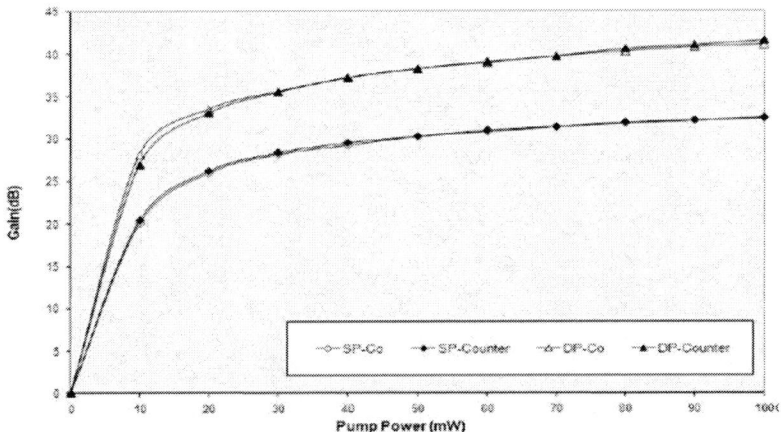

Figure 31. Gain vs. pump power for SP-Co, SP-Counter, DP-Co and DP-Counter configurations at small input signal power ($P_s = -50$ dBm).

The corresponding noise figure-pump power curves for SP-Co, SP-Counter, DP-Co and DP-Counter configurations are illustrated in Figure 32. As the pump power increases, the general trend shows a slight decrease in noise figure. This is anticipated given the fact that higher pump powers lead to a greater and more complete inversion in the EDFA which, in turn, translates to higher gains and lower noise figures.

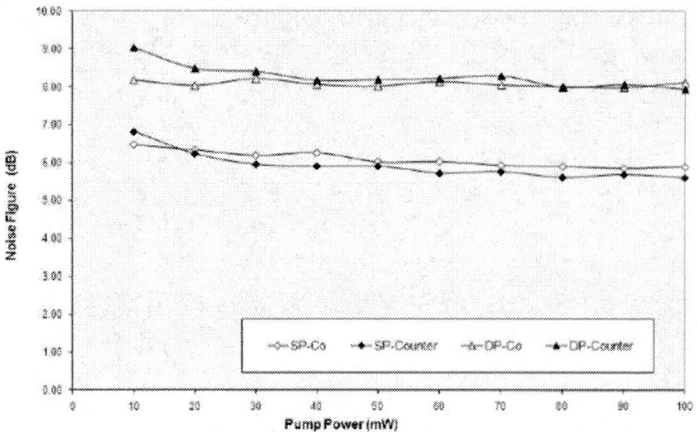

Figure 32. Noise figure vs. pump power for SP-Co, SP-Counter, DP-Co and DP-Counter configurations at small input signal power ($P_s = -50$ dBm).

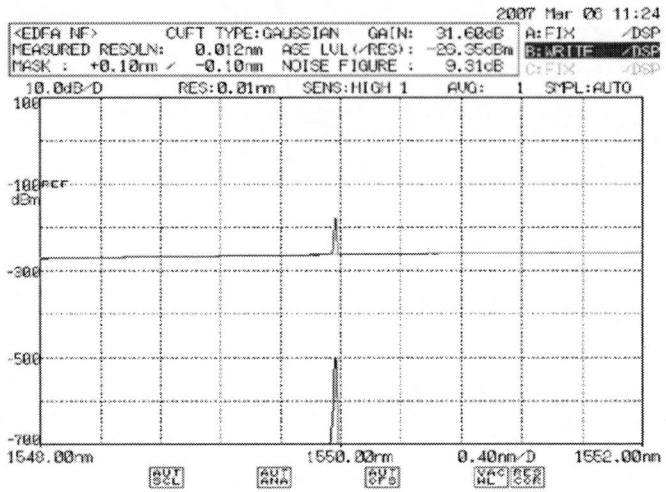

Figure 33. A typical OSA measurement display for DP-Co.

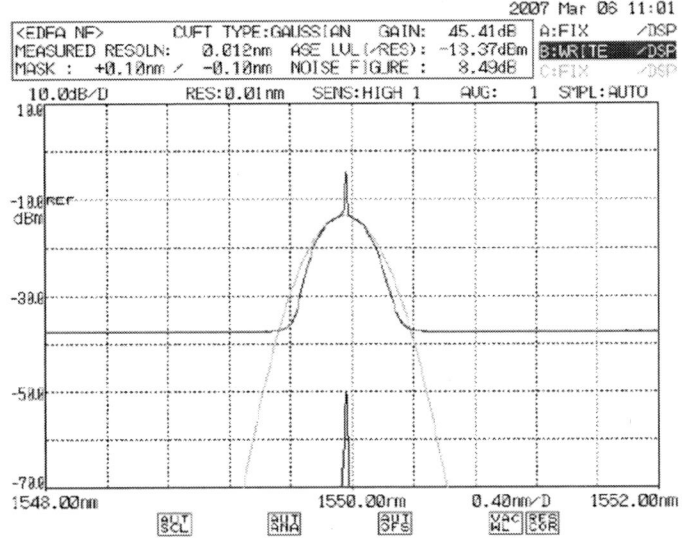

Figure 34. A typical OSA measurement display for DP-Co-TBF.

It is known that, for 1480 nm pumping, the fractional population in the metastable state increases with higher pump power, and is almost flat across the entire fiber length [44]. However, this is not true for the case of 980 nm pumping. As anticipated, the noise figure for the double-pass configuration is higher than that for the single-pass, given the fact that the ASE generated in the first pass is further amplified when it is re-circulated back into the amplifying medium during the second pass. Interestingly enough, pumping schemes have no appreciable effect on increasing the gain or reducing the noise figure for a given system configuration.

Unlike other configurations already discussed, the gain-pump power curves and the corresponding noise figure-pump power curves at a fixed signal wavelength λ_s = 1550 nm and input signal power of −50 dBm for DP-Co-TBF and DP-Counter-TBF configurations are unable to be carried out due to the presence of laser oscillations that set in at relatively low pump powers (~ 20 mW). These laser oscillations are observed as prominent spikes on the OSA display. With the use of TBF, the broadband ASE (which forms the optical noise) is greatly filtered out, allowing thereby the gain to grow unimpeded. This growth of the gain, coupled with the presence of Fresnel reflections at the interface of the splice of the EDFA and the standard single-mode fibre, create a laser cavity forming thereby laser oscillations described above. However, these laser oscillations can be suppressed in higher pump powers (~ 60 mW) with the use of higher input signal powers (~ 34 dBm).

A typical OSA display for the measurement of the performance characteristics for DP-Co and DP-Co-TBF configurations are shown in Figures 33 and 34, respectively. The behavior of gain and noise figure with respect to input signal power for SP-Co, SP-Counter, DP-Co, DP-Counter, DP-Co-TBF and DP-Counter-TBF configurations are illustrated in Figures 35 and 36, respectively. With the exception of DP-Co-TBF and DP-Counter-TBF, at input signal powers smaller than −30 dBm, the gain is constant, and this region is referred to as the small signal gain range. At input signal powers above −30 dBm, however, the gain starts to decrease as the amplifier saturates. It is especially noteworthy that, as revealed by Figure 35, at high input signal powers (e.g. −10 dBm), the three system configurations become indistinguishable, and converge to a common gain. It is only at small input powers that the different system configurations play a decisive role, and exhibit a difference in signal power amplification capability. Likewise, the noise figure remains constant in the small signal gain range, but shoots up dramatically as the amplifier saturates (Figure 36). As anticipated, implementation of an optical filter (like TBF) shows improvements in gain and noise figure of DP-Co-TBF and DP-Counter-TBF configurations over DP-Co and DP-Counter configurations, respectively.

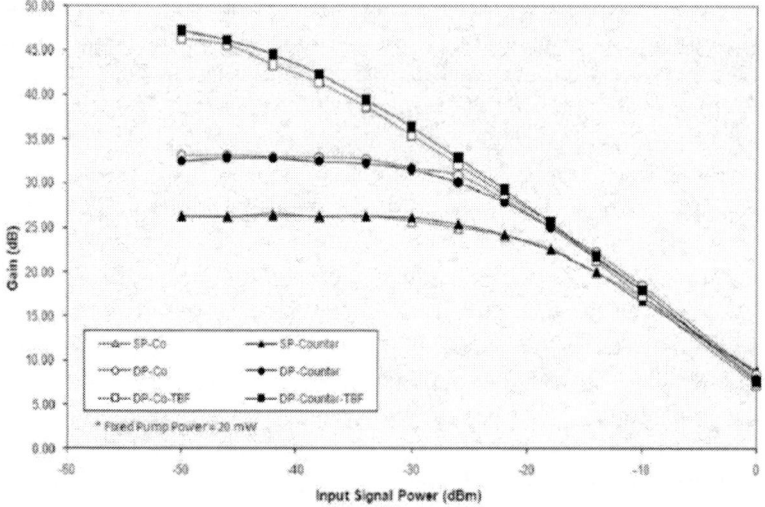

Figure 35. Gain vs. input signal power for various system configurations and pumping schemes at a fixed pump power of 20 mW.

Unlike the cases of single-pass or double-pass configurations, the gain features observed in DP-Co-TBF and DP-Counter-TBF configurations do not yield constant characteristics over the small signal gain regime of the input signal power. Instead, there is a gradual decrease in gain as the input signal power increases. Likewise, a gradual rise in noise figure is noticed as the input signal power increases. At the small signal regime ($\sim$ −50 dBm) and above medium pump powers ($\geq$ 30 mW), the ASE is filtered out to such an extent that the gain grows unimpeded much so that laser oscillations set in. The gain under these conditions cannot be measured accurately. When a low pump power (say 20 mW) is employed, gain saturation sets in fast because only a partial population inversion is created in the fiber. As such, even a small increase in input signal power causes the gain to drop rather abruptly, and the plateau region, present in the small signal regime in other configurations, remains absent (Figure 35).

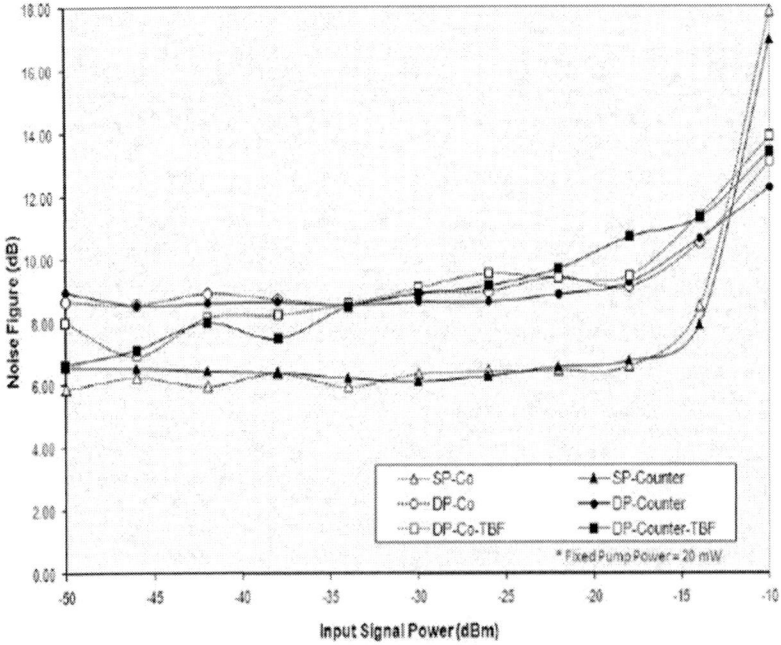

Figure 36. Noise figure vs. input signal power for various system configurations and pumping schemes at a fixed pump power of 20 mW.

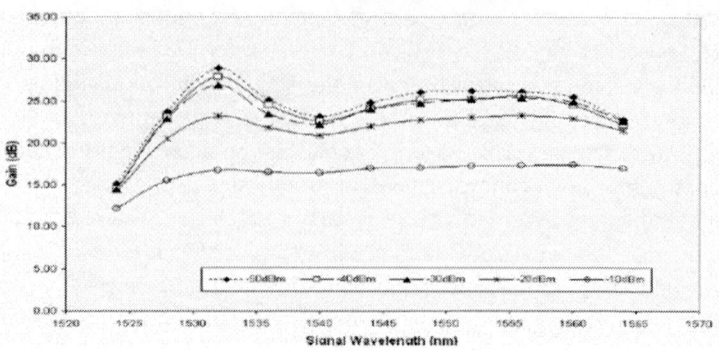

Figure 37. Gain spectra at a fixed pump power (20 mW) for various input signal powers (SP-Co case).

As stated above, the EDFA gain is wavelength-dependent. The non-uniformity of gain spectrum is due to the intrinsic material properties of silica glass fiber, namely the emission and the absorption cross-sections, which are functions of the signal and pump wavelengths [13]. The gain spectrum and the corresponding noise figure-wavelength characteristics for the SP-Co configuration are illustrated in Figures 37 and 38, respectively. An important point to note in this context is that, at a high input signal power (~ −10 dBm), the gain spectrum is smoothened out (i.e. flat) across the wavelength, and the noise figure is also much higher compared to those generated by a low-power input signal (~ −50 dBm).

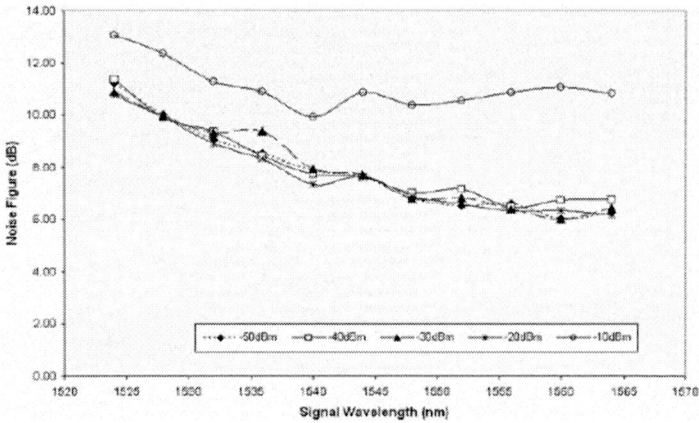

Figure 38. Noise figure vs. signal wavelength at a fixed pump power (20 mW) for various input signal powers (SP-Co case).

The gain spectrum and the corresponding noise figure-wavelength characteristics for various combinations of high and low pump powers with small and large input signal powers for SP-Co configuration are depicted in Figures 39 and 40, respectively. Based on Figure 39, two important observations can be made as follows:

i. For a small input signal power (−50 dBm), the characteristic non-uniform profile of gain is observed, regardless of whether a low (20 mW) or high (60 mW) pump power is utilized.
ii. A large input signal power (−10 dBm) causes the gain spectrum profile to be flattened out across the wavelength for both low (20 mW) and high (60 mW) pump powers.

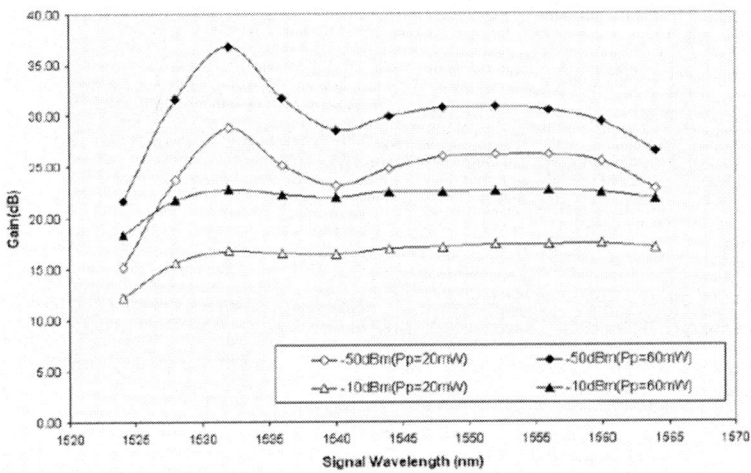

Figure 39. Gain vs. signal wavelength for various input signal and pump powers (SP-Co case).

Hence, it can be deduced that the gain spectrum profile is influenced only by the input signal power level, and is independent of the pump power utilized. Pump power does not affect the profile or the shape of the gain-wavelength curve. Instead, it merely serves to translate the entire profile up or down, depending on whether a high or low pump power is employed.

Based on Figure 40, the noise figure for small input signal power (−50 dBm) is smaller than that for a large input signal power (−10 dBm) for any given wavelength across the C-band. It also shows that, for a given fixed input signal power, the noise figure can be considered constant, within experimental error, regardless of whether a high or low pump power is employed.

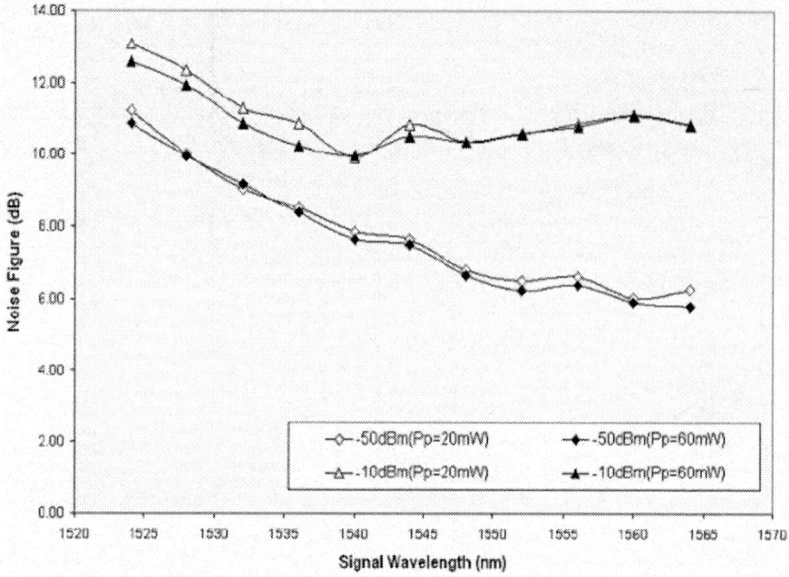

Figure 40. Noise figure vs. signal wavelength for various input signal and pump powers (SP-Co case).

3.3. COMPARISON OF THE PERFORMANCE CHARACTERISTICS UNDER DIFFERENT SITUATIONS

This section is devoted to a comparative study of all the accumulated experimental data for gain and noise figure characteristics with respect to signal wavelength corresponding to various system configurations and pumping schemes. This feature is particularly very important in view of dense wavelength division multiplexing (DWDM) systems. This is because of the fact that EDFAs amplify signals across the entire C-band simultaneously, and the gain spectrum determines how the individual channels (or wavelengths) are amplified [5]. This comparison remains essential as it helps to identify the acceptable optical circuits that yield the optimum gain and noise figure. Three different system configurations, namely single-pass, double-pass, and double-pass with TBF are considered [45]. Also, two different types of pumping schemes, viz. co- and counter-pumpings are compared side by side. The design parameters for all six different setups are also fixed whereby the length of EDFA is taken to be 14 m and the fiber has Er^{3+} ion concentration of 440 ppm. Likewise, the input signal power and pump power are also maintained at −50 dBm and 10 mW, respectively.

Figure 41 illustrates the gain spectrum for single-pass, double-pass, and double-pass with TBF configurations under the co- and the counter-pumping schemes. The meanings of all abbreviations used to indicate different configurations of optical circuits were stated before. We observe in Figure 41 that, for a given configuration, there is hardly any significant difference in gain characteristics corresponding to the different pumping schemes. However, the highest signal gain for small signal is obtained corresponding to the case of double-pass with TBF, regardless of the pumping scheme. As compared to the case of single-pass configuration, upon implementing a TBF in the system, an average of 14 dB improvement is observed across the signal wavelength range at the input signal power of −50 dBm. This is due to the efficient filtering of broadband ASE produced during the first pass amplification before it enters the second pass. The presence of ASE effectively reduces the signal gain, and this is best observed by comparing the gain characteristics obtained corresponding to the cases of double-pass with TBF and the single-pass configurations. The use of a TBF to filter out the unwanted ASE ensures less wastage of energy, improving thereby the signal gain.

The significant effect of gain robbing by the ASE is best illustrated in the case of double-pass configuration without TBF, where the buildup of ASE is so severe in short wavelength range (1528 nm − 1540 nm) that the signal gain (in this range) is even smaller compared to the case of single-pass configuration. This is due to the fact that the gain in an EDFA, which is wavelength-dependent, does not accumulate linearly from one pass to another. Instead, the gain is gradually removed from shorter wavelengths, and made available at the higher wavelengths, resulting in a dramatic change in gain spectrum and bandwidth loss. This observation is further supported by ref. [46], where the authors reported that, in a system of thirteen concatenated EDFAs, the usable bandwidth is shrunk to several nanometers. With the use of TBF in the system, large ASE buildup at short wavelength range is dramatically reduced. However, the characteristic profile of the gain spectrum and the usable bandwidth remain unaltered, and a corresponding increase in gain by an average of 14 dB across the C-band is observed.

Figure 42 illustrates the noise figure characteristics for the EDFA under consideration. It is observed that the double-pass arrangement introduces additional noise as compared to the situation corresponding to the single-pass one. This increase in noise figure is attributed to the fact that the developed noise due to ASE from the first pass is recycled once again through the amplifier during the second pass. It is further observed that the noise figure in the small signal regime corresponding to the double-pass scheme with TBF configuration is slightly lower than that obtained for the one without implementing TBF. This is essentially because the ASE from the first pass is filtered out by the TBF. In the case of double-pass without TBF, the ASE from the first pass is reflected, and amplified in the second pass. It is, however, to be noted at this point that the achieved signal gain essentially depends on the length of EDFA used in the experiment. In our case, we implemented an EDFA of 14 m in length; an increase in EDFA length would result in a lesser amount of gain for the same type of EDFA as the Er^{3+} ion doping concentration also plays a vital role in this regard.

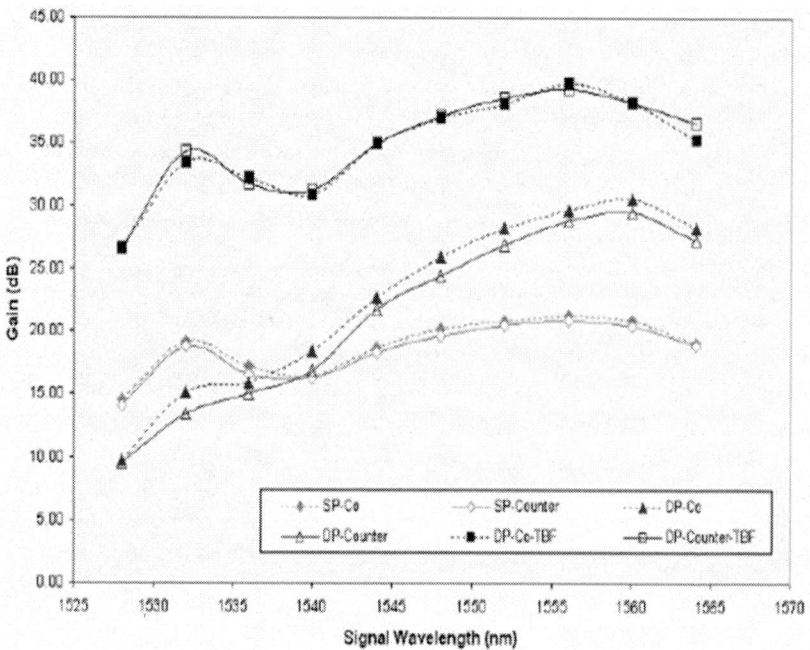

Figure 41. Gain vs. signal wavelength for various system configurations and pumping schemes at small input signal power (−50 dBm) and a fixed pump power (10 mW).

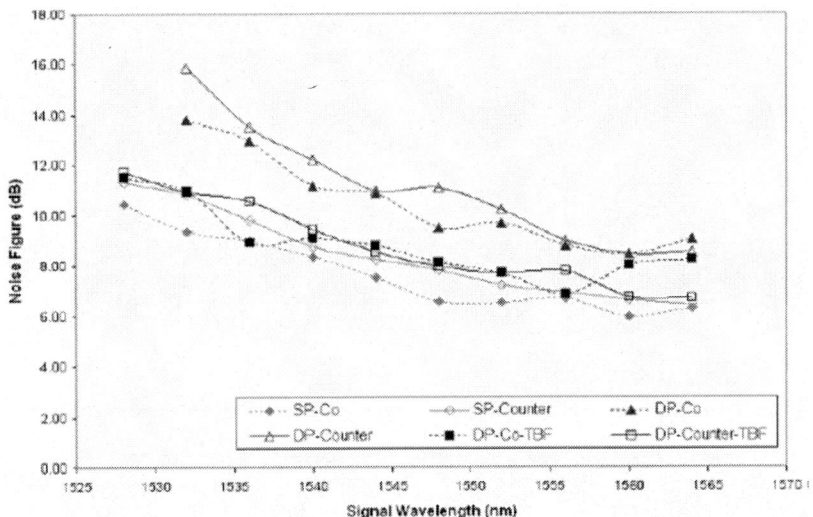

Figure 42. Noise figure vs. signal wavelength for various system configurations and pumping schemes at small input signal power (−50 dBm) and a fixed pump power (10 mW).

3.4. COMPARISON BETWEEN SIMULATION AND EXPERIMENTAL DATA

In this section, we compare the simulation data obtained from modeling and the experimental data corresponding to the Sp-Co configuration. Figures 43 and 44 present such a comparison of the gain spectrum and the corresponding noise figure characteristics for various combinations of high and low input signal power and pump power, respectively. The profiles of simulated and experimental gain spectrum curves exhibit similar general trends although displaying greater deviation at high pump power and low input signal power. The profiles of simulated and experimental noise figure characteristics curves too exhibit similar general trend although the simulation results show a very much lower noise figure in comparison to the corresponding experimental results. This is partly attributed to the fact that, while performing simulations, we ignored losses due to splices and irregularities in components which could cause Fresnel reflections; these parameters would present appreciable impacts in real experimental setups, forming thereby a contributing factor to a higher noise figure.

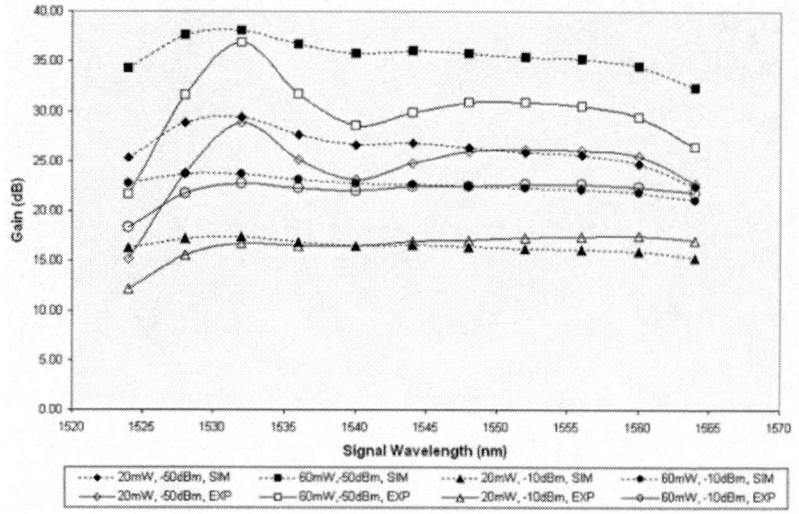

Figure 43. Simulation (SIM) and experimental (EXP) gain spectra corresponding to the SP-Co configuration for various values of small and large pump powers and input signal powers.

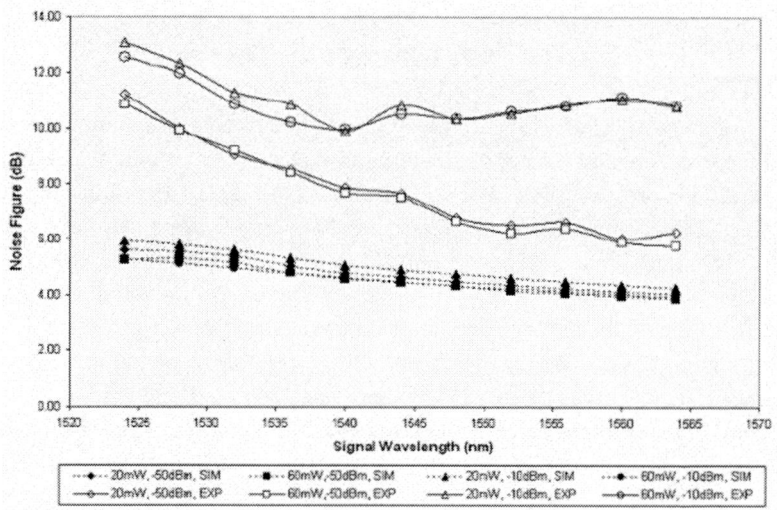

Figure 44. Simulation (SIM) and experimental (EXP) noise figure vs. signal wavelength corresponding to the SP-Co configuration for various values of small and large pump and input signal powers.

Chapter 4

SUMMARY AND CONCLUSION

EDFAs can be operated efficiently at pump wavelengths of both 980 nm and 1480 nm, and provide several advantages over regenerative repeaters and other amplification systems. These are thus recognized as all-optical amplifiers – the major catalyst in realizing the full strength of fiber-optic communication networks. The gain of EDFAs depend on pump power, signal wavelength, fiber length and doping concentration of Er^{3+} ions – the factors that can be optimized for various applications.

In this article, the operational principle of EDFA is discussed. Next, a comprehensive account of numerical modeling of EDFA is presented based on an established theoretical model developed earlier by Giles and Desurvire [2]. The model consists of a set of rate equations and propagation equations which form a set of coupled differential equations. The rate equations describe the temporal population distribution of Er^{3+} ions in ground, metastable and pump levels. At the steady state, populations in each of these energy levels remain fixed and constant. The propagation equations, in turn, represent the spatial evolutions of signal, pump and ASE powers across the longitudinal length of EDFA. Given that the coupled differential equations are non-linear in nature, they cannot be solved analytically. As such, a numerical technique implementing the RK4 method is resorted to carry out the numerical integration. Concurrently, the relaxation method is implemented to ensure the convergence of the solution. The simulation is then executed using an appropriate algorithm. Utilizing both sets of equations, the power corresponding to the signal, the pump and the ASE, anywhere further down the longitudinal length of EDFA can be determined. Therefore, for a given set of design parameters (e.g. input pump power, input signal power, Er^{3+} ion concentration and fiber length), gain and noise figure – the vital parameters in respect of the EDFA performance characteristics – can be evaluated. The simulated characteristics of gain and noise figure with respect to pump power, input signal power and input signal wavelength for SP-Co configuration are found to have close profile similarity with those previously reported by other investigators. In addition, the simulation results also show a good agreement, in terms of profile, with the experimental results obtained corresponding to the case of SP-Co EDFA configuration.

Subsequently, the experimental results for gain-pump power for all configurations except DP-Co-TBF and DP-Counter-TBF show a similar profile, i.e. the initial increase of gain with pump power, and then level off when the gain saturation is achieved. These results are useful in determining the minimum pump power needed to produce a complete population inversion in a given fiber, and hence, to achieve gain saturation. Pumping beyond this minimum pump power is merely wastage of energy. As for the DP-Co-TBF and the DP-Counter-TBF configurations, the presence of laser oscillations at high pump power and small input signal power make it impossible to measure the gain accurately. Besides that, the knowledge of the region where an amplifier can be operated efficiently remains equally important – this is obtained from the gain-input signal power characteristic. For all system configurations, it is shown experimentally that the amplifier should be operated in the small signal regime where the input signal power is small (≤ 30 dBm) so that it produces the best possible gain. Operating in the large signal domain reduces the efficiency of EDFA, since the ratio of the output to the input signal powers is approximately close to unity. Furthermore, in practice, amplifying a strong signal remains pointless.

Lastly, from the application point of view, gain spectrum is of great importance as it is wavelength-dependent, and hence, does not remain constant across a range of wavelengths. Given that EDFAs can amplify multiple wavelengths simultaneously in the C-band, this property is exploited in DWDM together with gain-flattening techniques to equalize gain across different wavelengths. Whenever a higher gain is needed, a double-pass configuration remains as the preferred choice over a single-pass configuration. From the experimental results, it is established conclusively that the DP-Co-TBF and the DP-Counter-TBF configurations not only produce better gain and lower noise figure, but also retain the original operational bandwidth, unlike the DP-Co and the DP-Counter configurations, wherein smaller wavelengths are severely attenuated, resulting in a loss of bandwidth.

REFERENCES

[1] http://www.aip.org/enews/physnews/1991/split/pnu020-3.htm.
[2] Desurvire, E; Giles, CR; Simpson, JR; Zyskind, JL. *Opt. Lett.*, 1989, 14, 1266−1268.
[3] Keiser, G. *Optical fiber communications*, McGraw Hill: Singapore, 2000.
[4] Kasap, SO. *Optoelectronic and photonics − Principles and practices*, Prentice Hall: Canada, 2001.
[5] Agrawal, G. *Fiber-optic communication systems*, Academic: New York, 2002.
[6] Hassami, A; Arzi, E; Seraji, FE. *Opt. Quantum. Electron.*, 2007, 39, 35−50.
[7] Dybdal, K; Bjerre, N; Pedersen, J.E; Larsen, CC. *Proc. SPIE*, 1989, 1171, 209−218.
[8] Ainslie, BJ; Craig, SP; Davey, ST; Wakefield, B. *Mater. Lett.*, 1988, 5, 139−144.
[9] Othonos, A; Wheeldon, J; Hubert, M. *Opt. Eng.*, 1995, 34, 3451−3455.
[10] Martin, JC. *Opt. Commun.*, 2001, 194, 331−339.
[11] Wysocki, PF; Simpson, JR; Lee, D. *IEEE Phot. Technol. Lett.*, 1994, 6, 1098−2000.
[12] Zech, H. *Opt. Fiber Technol.*, 1995, 1, 327−330.
[13] Quimby, RS. *Photonics and lasers: An introduction*, Wiley-Interscience, New York, 2006.
[14] Vengsarkar, AM; Lemaire, PJ; Judkins, JB; Bhatia, V; Erdogan, T; Sipe, JE. *J. Lightwave Technol.*, 1996, 14, 58−65.
[15] Wysocki, PF; Judkins, JB; Espindola, RP; Andrejco, M; Vengsarkar, AM. *IEEE Phot. Technol. Lett.*, 1997, 9, 1343−1345.

[16] Tachibana, M; Laming, RI; Morkel, PR; Payne, DN. *IEEE Phot. Technol. Lett.*, 1991, 13, 118−120.
[17] Su, SF; Olshansky, R; Joyce, G; Smith, DA; Baran, JE. *IEEE Phot. Technol. Lett.*, 1992, 4, 69−271.
[18] Inoue, K; Kominato, T; Toba, H. *IEEE Phot. Technol. Lett.*, 1991, 3, 718−720.
[19] Pan, JY; Ali, MA; Elrefaie, AF; Wagner, RE. *IEEE Phot. Technol. Lett.*, 1995, 7, 1501−1503.
[20] Yamada, M; Shimizu, M; Horiguchi, M; Okayasu, M; Sugita, E. *IEEE Phot. Technol. Lett.*, 1990, 2, 656−658.
[21] Bayart, D; Clesca, B; Hamon, L; Beylat, JL. *IEEE Phot. Technol. Lett.*, 1994, 6, 613−615.
[22] Yamada, M; Kanamori, T; Terunuma, Y; Oikawa, K; Shimizu M; Sudo, S; Sagawa, K. *IEEE Phot. Technol. Lett.*, 1996, 8, 882−884.
[23] Miniscalco, WJ. In *Rare earth doped fiber lasers and amplifiers*, Digonnet, MJF (Ed.), Dekker: New York, 1993.
[24] Bouzid, B; Ali, MB; Abdullah, MK. *IEEE Phot. Techol. Lett.*, 2003, 5 1195−1197.
[25] Naji, AW; Abidin, MSZ; Al-Mansoori, MH; Iqbal, SJ; Abdullah, MK; Mahdi, MA. *J. Opt. Commun.*, 2006, 27, 201−203.
[26] Pathmanathan, SS; Abdul-Rashid, HA; Choudhury, PK. *Asian J. Phys.*, 2008, 17, 241−244.
[27] Pathmanathan, SS; Abdul-Rashid, HA; Choudhury, PK. *Int. J. Microw. Opt. Technol.*, 2008, 3, 134−138.
[28] Urquhart, P. *Proc. IEE*, 1988, 135, 385−407.
[29] Saleh, AAM; Jopson, RM; Evankow, JD; Aspell, J. *IEEE Phot. Tech. Lett.*, 1990, 2, 714−716.
[30] Giles, CR; Desurvire, E. *J. Light. Tech.*, 1991, 9, 147−154.
[31] Yamashita, S; Okoshi, T. *IEEE Phot. Tech. Lett.*, 1992, 4, 1276−1278.
[32] Yu, A; O'Mahoni, MJ; Siddiqui, AS. *IEEE Phot. Tech. Lett.*, 1993, 5, 773−775.
[33] Hwang, S; Song, KW; Kwon, HJ; Koh, J; Oh, YJ; Cho, K. *IEEE Phot. Tech. Lett.*, 2001, 13, 1289−1291.
[34] Yi, LL; Zhan, L; Ji, JH; Ye, QH; Xia, YX. *IEEE Phot. Tech. Lett.*, 2004, 16, 1005−1007.
[35] Thyagarajan, K; Kakkar, C. *IEEE Phot. Tech. Lett.*, 2004, 16, 2448−2450.
[36] Zill, DG. *Differential equations with boundary-value problems*, Brooks/Cole Cengage Learning: Australia, 2009.
[37] Gilat, A. *MATLAB: An introduction with applications*, Wiley: NJ, 2005.

[38] Liu, JM. *Photonic devices*, Cambridge: New York, 2005.
[39] Zervas, MN; Leming, RI; Payne, DN. *J. Quantum Electron.*, 1995, 31, 472−480.
[40] Pedersen, B; Bjarklev, A; Povlsen, JH; Dybdal, K; Larsen, CC. *J Light. Tech.*, 1991, 9, 1105−1112.
[41] Hossain, N. *Modeling of hybrid EDFA/DRA for long haul optical fiber communication system*, M.Eng.Sc. Thesis, MMU, Cyberjaya, 2007.
[42] Ghatak, AK; Thyagarajan, K. *Introduction to fiber optics*, Cambridge: USA, 1998.
[43] Pathmanathan, S. *Performance characteristics of EDFAs in various system configurations*, M.Eng.Sc. Thesis, MMU, Cyberjaya, 2010.
[44] Becker, PC; Olsson, NA; Simpson, JR. *Erbium-doped fiber amplifiers: Fundamental and technology*, Academic: New York, 2002.
[45] Pathmanathan, SS; Md.-Yassin, SZ; Abdul-Rashid, HA; Choudhury, PK. *Optik*, 2010, 121, 184−187.
[46] Willner, AE; Xie, Y. In *Fiber optics handbook − Fiber devices and systems for optical communications*, Bass, M; Van Stryland, EW (Eds.), McGraw-Hill, New York, 13.1−13.31, 2002.

INDEX

A

algorithm, 49
amplification of light, vii
applications, 49, 52
AT&T, 5
Australia, 52

B

backward integration, 20
bandwidth, 1, 7, 36, 37, 45, 50
behavior, 24, 37, 40
boundary value problem, vii
broadband, 36, 39, 45
building blocks, 8

C

Canada, 51
catalyst, 49
communication, vii, 1, 2, 5, 6, 7, 49, 51, 53
communication systems, vii, 2, 5, 6, 7, 51
components, 7, 33, 47
computer simulations, 8
concentrates, 7
concentration, 5, 6, 8, 11, 19, 22, 23, 30, 36, 45, 46, 49

configuration, viii, 9, 21, 22, 23, 24, 25, 26, 27, 28, 29, 33, 34, 35, 36, 39, 42, 43, 45, 46, 47, 48, 50
convergence, 49
conversion, 19, 22

D

decay, 3, 4, 6, 15
degradation, 1, 5
density, 18, 19, 22, 26
derivatives, 13
differential equations, 13, 49
diode laser, 6
distribution, 6, 10, 29, 30, 31, 49
division, 5, 36, 44
doping, 1, 46, 49

E

earth, vii, 1, 52
electromagnetic, 5
electron, 3
electrons, 3, 4
emission, vii, 3, 4, 10, 11, 16, 17, 19, 20, 23, 25, 26, 27, 42
energy, 3, 4, 10, 13, 16, 29, 37, 45, 49, 50
equilibrium, 3, 13
erbium, vii, 1, 3, 5, 6, 9
evolution, 10, 16

excitation, 3

F

fabrication, 7, 9
fiber optics, 53
fibers, 19
filters, 7

G

generation, 5
groups, 19
growth, vii, 5, 15, 17, 25, 39

H

host, 19, 22, 25
hybrid, 53

I

ideal, 12, 14, 18
implementation, 7, 40
infinite, 18
integration, 29, 49
interface, 39
inversion, 4, 6, 14, 16, 17, 23, 26, 27, 29, 38, 41, 50
ions, 7, 10, 11, 15, 16, 23, 25, 26, 29, 49

L

lanthanide, vii, 1
lasers, 1, 5, 51, 52
lifetime, 3, 4
likelihood, 25
long distance, 1, 2

M

manufacturer, 3, 19
mathematics, 9
matrix, 15
measurement, 38, 39, 40
model, 2, 3, 8, 9, 10, 12, 19, 22, 49

modeling, vii, 2, 9, 10, 12, 14, 18, 19, 47, 49
models, 8, 9
multiplication, 16

N

nanometers, 45
network, 1, 5
noise, vii, viii, 5, 7, 14, 22, 23, 24, 26, 27, 29, 33, 36, 38, 39, 40, 41, 42, 43, 44, 46, 47, 48, 49, 50
non-linear rate, vii
numerical aperture, 6

O

observations, 43
optical communications, 53
optical fiber, 1, 5, 53
optical gain, 2, 14
optimization, 7
overlap, 11

P

parameter, 19
parameters, vii, 5, 7, 11, 12, 14, 19, 25, 36, 45, 47, 49
performance, vii, viii, 5, 7, 8, 9, 12, 19, 36, 40, 50
photonics, vii, 1, 51
photons, 3, 4, 15, 16, 17, 23, 25, 26
physics, 9
population, 3, 4, 10, 13, 14, 16, 17, 23, 26, 27, 29, 39, 41, 49, 50
power, vii, 1, 2, 5, 6, 7, 11, 12, 16, 21, 22, 23, 24, 25, 26, 27, 29, 30, 31, 36, 37, 38, 39, 40, 41, 42, 43, 45, 46, 47, 49, 50
probability, 25, 26
propagation, vii, 10, 13, 16, 18, 49
properties, 25, 42
pumping schemes, viii, 2, 8, 34, 39, 40, 41, 44, 45, 46, 47
pumps, 6

R

radius, 6
range, 9, 23, 40, 45, 50
rare-earth element, vii, 1
recommendations, iv
region, 5, 16, 17, 23, 37, 40, 41, 50
relaxation, vii, 19, 49
resolution, 29

S

saturation, 5, 23, 29, 37, 41, 50
scattering, 15
selecting, 22
semiconductor, vii, 1
shape, 43
signal quality, 5
signals, vii, 7, 37, 44
silica, 9, 15, 42
silicon, vii, 1
simulation, 2, 8, 14, 15, 16, 17, 19, 22, 23, 24, 29, 47, 49
Singapore, 51
software, 9, 14

specifications, 36
spectrum, 5, 19, 24, 25, 26, 27, 29, 36, 42, 43, 44, 45, 47, 50
speed, 2

T

telecommunications, 5
threshold, vii, 2, 36
transition, 3
transmission, 1, 5

U

uniform, 6, 7, 43

V

vacuum, 11
variations, 24

W

waste, 37
wavelengths, 19, 25, 37, 42, 44, 45, 49, 50